4th Natural Logic Meets Machine Learning Workshop (NALOMA23)

Nancy, France
23 June 2023

ISBN: 978-1-7138-9057-7

Printed from e-media with permission by:

Curran Associates, Inc.
57 Morehouse Lane
Red Hook, NY 12571

Association for Computational Linguistics
209 N. Eighth Street
Stroudsburg, Pennsylvania 18360

Phone: 1-570-476-8006
Fax: 1-570-476-0860

acl@aclweb.org

Additional copies of this publication are available from:

Curran Associates, Inc.
57 Morehouse Lane
Red Hook, NY 12571 USA
Phone: 845-758-0400
Fax: 845-758-2633
Email: curran@proceedings.com
Web: www.proceedings.com

IWCS2023

Proceedings of the 4th Natural Logic Meets Machine Learning Workshop (NALOMA23)

June 23, 2023

Association for Computational Linguistics (ACL)
209 N. Eighth Street
Stroudsburg, PA 18360
USA
Tel: +1-570-476-8006
Fax: +1-570-476-0860
acl@aclweb.org

Preface

This volume consists of papers presented at the fourth workshop entitled NAtural LOgic Meets MAchine Learning (NALOMA). The workshop was held physically at the 15th International Conference on Computational Semantics (IWCS) in 2023, in Nancy, France.

NALOMA aims to bridge the gap between machine learning/deep learning and symbolic/logic-based approaches to Natural Language Inference (NLI), and it is one of the only workshops organized to do so. The workshop also lays focus on theoretical notions of NLI which influence the way approaches to NLI can and should operate.

We thank everyone who submitted papers to the meeting, including the authors who submitted nonarchival extended abstracts that are not part of these proceedings. The meetings were enriched by the inspiring talks of our invited speakers: Claire Gardent and Johan Bos. We also thank all and everyone who served on the program committee (most served twice): Lasha Abzianidze, Katrin Erk, Hai Hu, Thomas Icard, Katerina Kalouli, Larry Moss, and Hitomi Yanaka. The meeting would not have been possible without the encouragement and organizational support that we received from the chairs of IWCS2023, Ellen Breitholtz and Maxime Amblard.

When people combine research communities, the intent is not merely to talk together but also to find joint intellectual projects. NALOMA's parents are logic and symbolic AI on one side, and machine learning on the other side. As we welcome NALOMA to its third year, we watch expectantly for those joint projects.

Stergios Chatzikyriakidis and Valeria de Paiva

Organizers:

Stergios Chatzikyriakidis (co-Chair), University of Crete

Valeria de Paiva (co-Chair), Topos Institute

Program Committee:

Aikaterini-Lida Kalouli, Ludwig Maximilian University of Munich

Lasha Abzianidze, Utrecht University

Katrin Erk, University of Texas at Austin

Hai Hu, Shanghai Jiao Tong University

Thomas Icard, Stanford University

Lawrence S. Moss, Indiana University

Hitomi Yanaka, University of Tokyo and Riken Institute

Invited Spekars:

Claire Gardent, Director of Research (first class), CNRS, LORIA, Nancy

Johan Bos, Professor of Computational Linguistics, University of Groningen

Table of Contents

Conference Program

Invited Talk by Claire Gardent: "What kind of errors are made by neural generation models and why?"

Evaluating Large Language Models with NeuBAROCO: Syllogistic Reasoning Ability and Human-like Biases
Risako Ando, Takanobu Morishita, Hirohiko Abe, Koji Mineshima and Mitsuhiro Okada

SpaceNLI: Evaluating the Consistency of Predicting Inferences In Space
Lasha Abzianidze, Joost Zwarts and Yoad Winter

Does ChatGPT Resemble Humans in Processing Implicatures?
Zhuang Qiu, Xufeng Duan and Zhenguang Cai

Recurrent Neural Network CCG Parser
Sora Tagami and Daisuke Bekki

Gijs Wijnholds: "Assessing Monotonicity Reasoning in Dutch through Natural Language Inference"

Daiki Matsuoka, Daisuke Bekki and Hitomi Yanaka: "Analyzing Relative Tense in Japanese with Dependent Type Semantics"

Invited Talk by Johan Bos: "Semantic Parsing with Large Language Models" (Chair: Staffan Larsson)

TTR at the SPA: Relating type-theoretical semantics to neural semantic pointers
Staffan Larsson, Robin Cooper, Jonathan Ginzburg and Andy Luecking

Triadic temporal representations and deformations
Tim Fernando

Discourse Representation Structure Parsing for Chinese
Chunliu Wang, Xiao Zhang and Johan Bos

Evaluating Large Language Models with NeuBAROCO: Syllogistic Reasoning Ability and Human-like Biases

Risako Ando
risakochaan@keio.jp

Takanobu Morishita
morishita@keio.jp

Hirohiko Abe
hirohiko-abe@keio.jp

Koji Mineshima
minesima@abelard.flet.keio.ac.jp

Mitsuhiro Okada
mitsu@abelard.flet.keio.ac.jp

Keio University

Abstract

This paper investigates whether current large language models exhibit biases in logical reasoning, similar to humans. Specifically, we focus on syllogistic reasoning, a well-studied form of inference in the cognitive science of human deduction. To facilitate our analysis, we introduce a dataset called NeuBAROCO, originally designed for psychological experiments that assess human logical abilities in syllogistic reasoning. The dataset consists of syllogistic inferences in both English and Japanese. We examine three types of biases observed in human syllogistic reasoning: belief biases, conversion errors, and atmosphere effects. Our findings demonstrate that current large language models struggle more with problems involving these three types of biases.

1 Introduction

Syllogistic inferences and their various variants have been extensively studied since *Prior Analytics* by Aristotle in the 4th century BC. While the Aristotelean syllogism is a small part of the predicate logic and a limited inference system when compared to, for example, the formal system of logical inference rules, there has been recently a revival movement of Aristotelean syllogism and its variants, including natural logic (van Benthem, 1986; Sánchez Valencia, 1991; Moss, 2015). This renewed attention arises from the perspective of viewing syllogistic inferences as a "natural" inference rule applicable to our everyday reasoning in ordinary language.

Not only is there a re-evaluation of the significance of syllogistic inferences and their variants in relation to their usefulness in ordinary language, but they are also considered as a benchmark for various inference studies in different disciplines. For example, cognitive psychological studies of logical inferences (Stenning and Van Lambalgen, 2012), diagrammatic logical inference studies (Sato and Mineshima, 2015), neuroscientific studies of logical inferences (Goel et al., 2000), all draw upon syllogistic reasoning as a point of reference. On the other hand, the recent developments of deep-learning-based AI-tools of natural languages, in particular, the state-of-the art Large Language Models (LLMs), including BERT (Devlin et al., 2019) and GPT (Brown et al., 2020), are remarkable. These tools hold the potential to be useful for logical inferences. However, there is still a need for further accumulation of studies on the use of AI models for logical inferences in natural language.

In this paper, we explore the potential of LLMs for performing logical inferences, with a specific focus on using syllogistic inferences as a benchmark. We present the NeuBAROCO dataset, a new dataset consisting of syllogistic inferences in both English and Japanese. The dataset is derived from the question collection BAROCO (Shikishima et al., 2009), which has been used in various studies in Japan to evaluate human syllogistic reasoning abilities.

The field of cognitive science of human reasoning has not yet been fully integrated with the recent advancements in AI, despite its potential to provide valuable insights for AI inference research. Considering the significant attention given to biases in studies of logical inference within cognitive psychology and cognitive science (Evans, 1989; Evans et al., 1993), our primary focus lies in assessing whether LLMs exhibit the biases observed in human logical inferences. We focus on three types of human biases that have been studied in cognitive science, namely, belief biases, conversion errors, and atmosphere effects (see Section 4). We explore the extent to which currently available LLMs for natural language inference can effectively address syllogistic inferences that are susceptible to errors resulting from these biases.

The contributions of the paper are summarized as follows.

1. We present the NeuBAROCO dataset, specif-

Proceedings of the 4th Natural Logic Meets Machine Learning Workshop (NALOMA23), pages 1–11
June 20, 2023.

ically designed for syllogistic inferences, which serves as a valuable resource for examining human-biases in language models.

2. Using this dataset, we evaluate the logical reasoning ability of several recent LLMs both for English and Japanese.

3. Our evaluation reveals that the current LLMs exhibit significant shortcomings when faced with problems that are prone to errors resulting from the three biases of interest.

The dataset will be made publicly available for research on human and machine understanding of logical inferences.

2 Background and Related Work

2.1 Syllogism

A syllogism consists of the following four types of categorical sentences described as **A**, **E**, **I**, and **O**.

Type	Form	Description
A	All S are P	Universal affirmative
E	No S are P	Universal negative
I	Some S are P	Particular affirmative
O	Some S are not P	Particular negative

Table 1: Four types of categorical sentences

Each syllogism is composed of two premises (**P1**, **P2**) and one conclusion (**C**). Thus, each type of syllogisms is identified by the types of three sentences. The following is an example of EIO-type syllogism (called *Ferio* in the traditional mnemonic), which is a valid pattern of syllogism.

P1: No B are C. (**E**-type)
P2: Some A are B. (**I**-type)
C: Some A are not C. (**O**-type)

As outlined in Section 1, syllogisms have been extensively studied in the fields of logic and psychology. To assess the logical reasoning capabilities of current language models in natural language inferences, we use a dataset that encompasses various types of syllogistic inferences.

2.2 Machine learning and logical inference

In recent years, syllogistic inferences and their variants have been used to examine the reasoning ability of machine learning based models for natural language inferences (NLI).

Richardson et al. (2020) examined the capacities of NLI models to perform various types of logical inferences that involve boolean operators, quantifiers, comparatives, conditionals, negation, and counting expressions. The study utilized synthetically generated data. Their findings indicate that models fine-tuned with NLI datasets perform well, suggesting that NLI models enhance their accuracy when provided with additional datasets as input. However, due to the artificial nature of the data, the majority of the inferences in each type of inferences share a similar structure, making it relatively easy for machine learning algorithms to solve such similar problems. In contrast, our study utilizes human-generated data, with a specific emphasis on inferences that elicit human-like biases. Furthermore, our study focuses on the capacity of LLMs to solve syllogistic inferences without requiring fine-tuning with NLI training examples.

Yanaka et al. (2019) examined monotoniticy inferences, those logical inferences that are licensed by substituting general terms in quantified sentences. A testset for monotonicity inference was created semi-automatically and then tested on various language models. One interesting finding is that all the models performed poorly on a class of inferences involving negative contexts (the so-called downward monotone inferences). Monotonicity inferences are structurally simpler than syllogistic inferences in that they consist of single-premises; by contrast, syllogisms are composed of multiple premises. Combination of quantifiers in syllogisms such as *no* and *some* can be more challenging than monotonicity inferences.

Schlegel et al. (2022) conducted an empirical study to investigate the detection of formally valid inference within controlled fragments of natural language. These fragments were specifically designed to increase the complexity of the satisfiability problem. In their study, each fragment consisted of artifically generated sets of English sentences that incorporated determiners such as *every*, *some*, *no*, negation *not*, and relative clauses, within the context of a vocabulary comprising count nouns and transitive verbs. The findings of Schlegel et al. (2022) indicate that transformer-based language models fine-tuned with training data tend to exhibit overfitting to superficial patterns present in the data, rather than acquiring the logical principles that govern reasoning within these fragments. In other words, the models seem to focus on surface-level features rather than grasping the underlying logical principles. Furthermore, according to Schlegel et al.

(2022), the ability of neural networks to learn and solve the various satisfiability problems does not appear to align with the complexity classes associated with the elicited fragments. Our study focuses on a small yet manually controlled dataset, as opposed to a large corpus of artificially generated data. Specifically, we manually annotate problems that are susceptible to human-like biases with corresponding labels. This approach facilitates meaningful comparisons between the syllogistic inference capabilities of humans and models.

Closest to our work is Dasgupta et al. (2022), which reveals that large language models show content effects (i.e., what we called *belief biases*) in syllogism reasoning as well as humans. They introduced new datasets of abstract logical inferences including syllogisms. Each syllogism is annotated with the information about whether or not a proposition is consistent with human beliefs and knowledge. They found that when the conclusion of an inference contradicts reality, the language model exhibits a strong bias towards classifying the argument as invalid, regardless of its logical validity. Our experimental results provide further support of these findings and demonstrate that similar effects are observed not only in English but also in Japanese, a typologically different language from English. Furthermore, we expand our focus beyond belief biases to include various types of biases such as conversion errors and atmosphere effects. We systematically examine the impact of these biases on LLMs with the ultimate objective of comparing the performance of LLMs in logical reasoning to that of humans.

3 The NeuBAROCO Dataset

3.1 Background: the BAROCO dataset

BAROCO is the collection of logical inference questions to examine subjects' ability of logical inference. The questions of BAROCO are mainly composed of syllogistic inferences and their variants. BAROCO has been used in various studies on human logical inference abilities. BAROCO was first used for behavioral genetic studies with the twin method in Shikishima et al. (2006, 2009), where the genetic factor and the environmental factor of the logical inference ability were measured. 500 twin pairs (1,000 participants) were asked to answer 100 questions about a version of BAROCO. The results were then compared with the subjects' scores on a standard IQ test that typically did not include logical inference abilities. Additionally, correlations were explored between logical inference abilities and decision-making skills in the fields of behavioral economics and cognitive sociology. For instance, Shikishima et al. (2015) investigated the relationship between logical inference abilities and Allais's decision-making task, along with other related studies.

3.2 Data construction

The full version of the original BAROCO dataset comprises a collection of 209 logical inferences divided into seven sections, each containing different types of questions. The version of BAROCO called "BAROCO-ALL" encompasses a total of 200 questions, which includes the following three sections. Examples of each section will be presented in Table 2.

(1) Abstract syllogism inferences: This section consists of inferences where the terms used in the sentences are represented by capital letters of the alphabet.

(2) Contentual (belief-consistent) syllogistic inferences: In this section, the inferences are constructed using concrete nouns commonly used in ordinary language.

(3) Belief-inconsistent syllogistic inferences: This section introduces inferences where belief-inconsistent sentences may appear within the inference itself.

Most questions in BAROCO dataset consist of two premises and three options for the correct answer. In the original setup, the participants were asked to choose one logically valid conclusion from the given options. We transformed each question into a format commonly used for evaluating NLI models, where inferences are categorized as *entailment*, *contradiction*, or *neither* (which we call *neutral*). In accordance with the format of syllogisms, each inference consists of two premises and one conclusion. We manually assigned each inference with the approapriate label of *entailment*, *contradiction*, and *neutral*.

The resulting dataset obtained from this process is referred to as NeuBAROCO. In total, there are 375 inference problems in the NeuBAROCO dataset, with 122 instances labelled as *entailment*, 71 instances labelled as *contradiction*, and 182 instances labelled as *neutral*.

Type	Language	Example
Symbol	English	**P1**: All A are B. **P2**: All B are C. **C**: All A are C.
	Japanese	**P1**: すべての A は B である **P2**: すべての B は C である **C**: すべての A は C である
Consistent	English	**P1**: One friend of Taro is a friend of Paul. **P2**: All of Paul's friends are German. **C**: One of Taro's friends is German.
	Japanese	**P1**: 太郎のある友人はポールの友人である。 **P2**: ポールのすべての友人はドイツ人である。 **C**: 太郎のある友人はドイツ人である。
Inconsistent	English	**P1**: Some animals are human beings. **P2**: All animals are tomatoes. **C**: Some humans are tomatoes.
	Japanese	**P1**: ある動物は人間である。 **P2**: すべての動物はトマトである。 **C**: ある人間はトマトである。

Table 2: Examples of symbolic, consistent, and inconsistent syllogism in the NeuBAROCO dataset. The English sentences (**P1**, **P2**, **C**) in each example correspond to the respective Japanese sentences. The correct label for all examples is *entailment*.

The BAROCO dataset was written in Japanese. We translated each problem into English using the DeepL translation tool (`https://www.deepl.com/translator`). We manually checked and adjusted the wording of each sentence, ensuring that they conform to the patterns of categorical sentences. We normalized the quantifiers in the English sentences. We used *all* or *every* for universal quantification in **A**-type sentences, and *some*, *a certain*, or *one of* for existential quantification in **I**-type and **O**-type sentences, and *no* for universal negative in **E**-type sentences. To prevent the sentences from being interpreted as generic statements, we refrained from using the indefinite article *a* (or *an*) for existential quantification. The presence of the indefinite article can lead to a generic interpretation, such as in the sentence *A cat is an animal*. This ensures consistency and clarity in the translation of the original Japanese sentences into English.

3.3 Annotation

We annotated each inference problem in the NeuBAROCO dataset as to what type of inference it is and whether the sentences appearing in it are consistent with beliefs.

3.3.1 Types of logical inferences

There are two types of inferences in the dataset: basic syllogisms and extended syllogisms.

Basic syllogisms As explained in Section 2, basic syllogisms consist of two premises (**P1**, **P2**) and one conclusion (**C**). We annotate each basic syllogism with the types of premises and conclusion. The following is an example of **IAI**-type syllogism:

P1: Some A are B. (**I**-type)
P2: All B are C. (**A**-type)
C: Some A are C. (**I**-type)

The first premise is **I**-type. The second premise is **A**-type. The conclusion is **I**-type. Therefore, the inference is labeled as **IAI**.

Extended syllogisms Extended syllogisms can be classified into two types. One is a boolean inference where conjunction *and* and *or* appear between terms. The following is an example:

P1: All A or B are C.
P2: No C are D.
C: No B are D.

The other is a hypothetical syllogism, one of whose premises is a conditional sentence of the form *If P then Q*. Here, *P* or *Q* can be a negated sentence. The following is an example:

P1: If Hanako has blood type O,
then Hanako's daughter has blood type B.
P2: Hanako's daughter does not have blood type B.
C: Hanako does not have blood type O.

In the dataset, there are 318 basic syllogisms and 57 extended syllogisms.

3.3.2 Belief consistency

We also classify the inferences into three distinct types based on the types of sentences they contain: *symbolic*, *consistent*, and *inconsistent*. Table 2 shows examples of each type.

Symbolic A symbolic inference is composed of sentences where all the terms are abstract symbols (alphabets). For humans, they can be considered to be neutral with respect to beliefs.

Consistent An inference is labelled as *consistent* if all of the premises and conclusion are consistent with common-sense beliefs. In the case of the example in Table 2, all the sentences, i.e., *One friend of Taro is a friend of Paul*, *All of Paul's friends are German*, and *One of Taro's friends is German*, can be interpreted consistent with belief.

Inconsistent An inference is labelled as *inconsistent* if at least one of of the premises and conclusion is inconsistent with common-sense beliefs, that is, it goes against what is commonly believed or accepted. In the case of the example in Table 2, the contents of two sentences, *Some humans are not living things* and *None of the animals are human* are contrary to common sense.

There are 95 instances of *symbolic*, 167 instances of *consistent*, and 102 instances of *inconsistent*. For cases where the judgment of belief consistency is unclear, we classify them as *others*. We encountered 11 instances that fell into this category.

4 Human-like Biases

Based on the above classification of the types of syllogistic inferences and sentences, we examine three types of human-like biases that can cause reasoning errors: belief biases, conversion errors, and atmosphere effects. We annotated information to each inference in the dataset to make explicit which inferences are misjudged by these biases.

4.1 Belief biases

Belief bias is one of the most well-known biases causing inference errors and has been applied to various types of logical inferences including syllogisms and Wason's selectional task (Evans, 1989; Newstead et al., 1992; Evans et al., 1993). It is widely recognized that people tend to have trouble in determining whether an inference is valid when it includes a sentence contrary to common sense. For example, the inference that is labelled as *inconsistent* in Table 2, repeated here, has inconsistent sentences **P2** and **C**:

P1: Some animals are human beings.
P2: All animals are tomatoes.
C: Some humans are tomatoes.

Although the correct label for this problem is *entailment*, the fact that the conclusion **C** is contrary to beliefs may lead some to judge it as *contradiction* instead of *entailment*, regardless of its logical validity. Similarly, in the following example, while **P2** is contrary to our beliefs, the conclusion **C** remains consistent. Hence, one might judge the inference as *entailment* rather than *neutral* due to the belief-consistency of the conclusion.

P1: All canines are animals.
P2: All animals are robots.
C: No canine is a robot.

As mentioned in Section 3.3.2, when either one of the premises or the conclusion is inconsistent with our beliefs, we assigned the *inconsistent* label to the inference. We investigate whether or not NLI models are influenced by this type of belief biases.

Conversion errors are errors in syllogisms caused by the incorrect interpretation of terms that appear in premises. There are at least two types of errors, called *illicit conversion* (Wilkins, 1928; Newstead, 1989; Geurts, 2003):

1. The tendency to interpret *All A are B* as equivalent to *All B are A* (**A**-type)

2. The tendency to interpret *Some A are not B* as equivalent to *Some B are not A* (**O**-type).

Note that *All A are B* and *Some A are not B* mean $A \subseteq B$ and $A \cap \overline{B} \neq \emptyset$, respectively, in the standard predicate logic[1], hence terms A and B are not convertible. Table 3 shows some examples of syllogistic inference problems where conversion errors cause wrong answer.

[1]In traditional syllogisms, **A**-type sentence *All A are B* implies that A is not empty. The BAROCO dataset follows this traditional interpretation (the *existential import* of universal expressions) when annotating the gold labels.

Type of syllogisms	Language	Example
AAA	English	**P1**: All B are A. **P2**: All B are C. **C**: All A are C.
	Japanese	**P1**: すべての B は A である **P2**: すべての B は C である **C**: すべての A は C である
AOO	English	**P1**: All chimpanzees are animals. **P2**: Some animals are not primates. **C**: Some primates are not chimpanzees.
	Japanese	**P1**: すべてのチンパンジーは動物である。 **P2**: ある動物は霊長類でない。 **C**: ある霊長類はチンパンジーでない。
OAO	English	**P1**: Some ghosts are not students. **P2**: All students are humans. **C**: Some humans are not ghosts.
	Japanese	**P1**: ある幽霊は生徒でない。 **P2**: すべての生徒は人間である。 **C**: ある人間は幽霊でない。
EAO	English	**P1**: No robot is human. **P2**: Every human being is a living organism. **C**: A certain living organism is not a robot.
	Japanese	**P1**: どのロボットも人間でない。 **P2**: すべての人間は生物である。 **C**: 生物のあるものはロボットでない。
AII	English	**P1**: All humans are animals. **P2**: Some robots are animals. **C**: Some humans are robots.
	Japanese	**P1**: すべての人間は動物である。 **P2**: あるロボットは動物である。 **C**: ある人間はロボットである。

Table 3: Examples of syllogistic inference problems where conversion errors give wrong answer. The correct label for all the examples is *neutral*. The English sentences (**P1**, **P2**, **C**) in each example correspond to the respective Japanese sentences.

We identified the syllogisms in which the correct answer is *neutral* and whose premises contain a sentence of the form *All A are B* or *Some A are not B*, and whose correct answer changes from *neutral* to *entailment* by applying conversion to either one or both premises. We annotate the *conversion* label to this type of problems. In the original dataset, there are only 10 such problems. Thus we expanded the dataset by adding more *conversion* problems that have the types not included in the original dataset. We fixed a set of schematic types to be added and obtained instances of these types by substituting abstract terms with concrete nouns in the dataset. In total, there are 70 problems to which the *conversion* label is assigned in the NeuBAROCO dataset.

4.2 Atmosphere effects

Atmosphere effects are one of the inferential biases that can be traced back to studies in the 1930s (Woodworth and Sells, 1935; Khemlani and Johnson-Laird, 2012). It can be interpreted as two principles (Chater and Oaksford, 1999):

1. *The principle of quality*: if one or both premises are negative (**E**-type or **O**-type), the conclusion should be negative; otherwise, it is positive (**A**-type or **I**-type).

2. *The principle of quantity*: if one or both premises are particular (**I**-type or **O**-type), then the conclusion will be particular; oth-

Type of syllogisms	Language	Example
AOE	English	**P1**: All animals are living things. **P2**: Some humans are not living things. **C**: None of the animals are human.
	Japanese	**P1**: すべての動物は生物である。 **P2**: ある人間は生物でない。 **C**: どの動物も人間でない。
OAI	English	**P1**: A certain police officer is not a public servant. **P2**: All human beings are public servants. **C**: Some police officer is a human being.
	Japanese	**P1**: ある警察官は公務員でない。 **P2**: すべての人間は公務員である。 **C**: ある警察官は人間である。

Table 4: Examples of syllogistic inference problems where atmosphere effects can give wrong answer. The correct label for all the examples is *neutral*. The English sentences (**P1**, **P2**, **C**) in each example correspond to the respective Japanese sentences.

erwise, it is universal (**A**-type or **E**-type).

Previous psychological experiments based on the original BAROCO data (for Japanese) have shown that **O**-type inference is particularly difficult for logically untrained human participants (Shikishima et al., 2009). Thus among various patterns, we focus on the cases where at least one premise is an **O**-type or **I**-type sentence. We assign the *atmosphere* label to an inference if its correct answer is *neutral* and (1) at least one of the premises is *O*-type and the conclusion is either **E**-type, **I**-type, or **O**-type or (2) at least one of the premises is **I**-type and the conclusion is either **I**-type or **O**-type.

Table 4 shows some representative examples of syllogisms satisfying these conditions. In total, there are 104 problems labelled as *atmosphere* in the dataset.

5 Experiments

5.1 Experimental settings

We evaluate syllogistic reasoning ability of deep neural networks, and in particular state-of-the-art large language models using our NeuBAROCO dataset. We evaluate transformer-based pre-trained language models, RoBERTa (Liu et al., 2019) and BART (Lewis et al., 2020), both being fine-tuned with the MultiNLI dataset (Williams et al., 2018). We use the models available in the `transformers` library, `roberta-large-mnli` and `facebook/bart-large-mnli`.[2] We also evaluate OpenAI's GPT-3.5 model, an improved version of GPT-3 (Brown et al., 2020). We use OpenAI's ChatGPT API with the `GPT-3.5-turbo` model.[3] Table 5 shows some prompt examples in English and Japanese. We chose the best one among several prompts we tried. When we use a phrase like *logical inference* or *syllogism* instead of *inference*, the performance became worse.

To test whether the models show belief biases, conversion errors, and atmosphere effects, we tested how well the models can answer correctly to the problems labelled as *inconsistent*, *conversion*, and *atmosphere*, and compared the accuracies with the total average on the NeuBAROCO dataset.

5.2 Results and discussion

5.2.1 Overall results

Table 6 shows the overall accuracy of each model and the accuracy on each correct label. Table 7 shows the accuracy of the models on basic and extended syllogisms.

RoBERTa The overall accuracy was low (34.67%). The accuracy for the *contradiction* problems was very high (74.65%), although the accuracy of the *neutral* problems was very low (0.55%). The accuracy of the *entailment* problems was between the two (62.3%). The accuracy on extended syllogism was higher than basic syllogism (54.39% and 31.13%).

BART The results on BART shows the same tendency. The overall accuracy was low (35.2%).

[2] https://github.com/huggingface/transformers

[3] https://platform.openai.com/docs/model-index-for-researchers

Carefully evaluate the following inference, and determine whether the premises entail or contradict the conclusion. Answer with entailment, contradiction, or neither.
Premise 1: Some A are B.
Premise 2: All B are C.
Conclusion: All A are C.

次の推論を注意深く読み、前提が結論を含意するか、矛盾するかを判定しなさい。「含意」「矛盾」「どちらでもない」のいずれかで答えなさい。
前提1：あるAはBである。
前提2：すべてのBはCである。
結論：すべてのAはCである。

Table 5: Prompt examples in English and Japanese

Language	Models	All	Entailment	Contradiction	Neutral
English	RoBERTa	34.67	62.30	74.65	0.55
	BART	35.20	55.74	83.10	2.75
	GPT-3.5	51.73	79.51	38.03	38.46
Japanese	GPT-3.5	48.27	80.33	54.93	24.18

Table 6: Accuracy (%) of the models on all the inference problems and each correct inference label.

Language	Models	Basic	Extended
English	RoBERTa	31.13	54.39
	BART	31.13	57.89
	GPT-3.5	51.57	52.63
Japanese	GPT-3.5	46.86	56.14

Table 7: Accuracy (%) of the models on basic and extended syllogisms.

Language	Models	All	Symbol	Consistent	Inconsistent	Conversion	Atmosphere
English	RoBERTa	34.67	24.21	46.11	22.55	0.0	0.0
	BART	35.20	34.74	45.51	15.69	1.43	0.96
	GPT-3.5	51.73	61.05	56.89	31.37	25.71	39.42
Japanese	GPT-3.5	48.27	55.79	56.29	25.49	21.43	22.12

Table 8: Accuracy (%) of the models on each type of syllogistic inferences and biases.

BART answered correctly in most *contradiction* cases, but less in *entailment* and still less in *neutral* (83.1%, 55.74% and 2.75%). The performance on extended syllogism was better than that on basic syllogism (57.89% and 31.13%).

GPT-3.5 The overall accuracy was 51.73%. The accuracy on the *entailment* problems was very high (79.51%), while that on the *contradiction* and *neutral* problems were low (38.03% and 38.46%). We found little difference between basic and extended syllogism (51.57% and 52.63%).

5.2.2 Results on problems concerning biases

Table 8 shows the accuracy of the models on each type of syllogistic inferences and biases.

Belief biases Among the three types of inferences, *symbol*, *consistent*, and *inconsistent*, the performance on the *inconsistent* cases was the lowest in every model. We found a significant difference among the models on which of the *symbol* or *consistent* cases they answered most accurately. GPT-3.5 performed better in *symbol* and *consistent* than *inconsistent* (61.05%, 56.89% and 31.37%). For RoBERTa, the percentage of correct answers to the *consistent* cases was higher than in *inconsistent* and *symbol* cases (46.11%, 22.55% and 24.21%). The percentages of correct responses of BART were, in order of highest to lowest, *consistent*, *symbol*, and *inconsistent* (45.51%, 34.74% and 15.69%).

Overall, the results show a tendency that GPT-3.5 outperforms both RoBERTa and BART models in symbolic reasoning. Also, GPT-3.5 performed equally well on *symbol* and *consistent* problems. This suggests that *symbol* and *consistent* are relatively easy to handle in that both types of problems are not contrary to beliefs. These results contrast with the results on RoBERTa, which performed almost equally on *symbol* and *inconsistent* problems.

Conversion error Regarding the *conversion* problems, all models exhibit low performance. A striking difference exists between RoBERTa and BART, on one hand, and GPT-3.5, on the other. While RoBERTa and BART hardly answered correctly (0% and 1.43%), GPT-3.5 performed better, answering correctly almost a quarter of the cases, which is about half of the overall average (25.71%). The results show that all the models performed poorly on the problems where incorrect responses are made by conversion errors.

Atmosphere effects We found that the performances of the models were notably lower in *atmosphere* cases. While RoBERTa and BART hardly provide correct answer to *atmosphere* cases (0% and 0.96%), GPT-3.5 performed better (39.42%).

5.2.3 Results on the Japanese GPT model

The results on the Japanese GPT model shows the strikingly same tendency as the English GPT models. One notable exception is the performance on the *contradiction* problems (see Table 6), where the Japanese GPT-3.5 model performed better than the English GPT-3.5 models. By contrast, the performance on *atmosphere* problems was worse than that of the English model.

6 Conclusion

In this paper, we investigated syllogistic reasoning ability of current large language models in focusing on human-like biases that have been studied in the context of cognitive science of human reasoning. The experiments indicated that the state-of-the-art models fail for problems where errors are caused by various human-like biases, and that there is large room for improvement in deductive reasoning capabilities of large language model.

Among other things, our results on conversion errors suggest the importance of distinguishing the problems of interpreting sentences (in particular, interpreting quantifiers and negation) from the problem of performing logical inferences. For conversion cases, there is abundant room for discussion on at which level the models mistook. Further inquiry into this issue could provide insight into a better understanding of the behavior of neural models.

There remain many issues to be addressed. First, although we tested the models with no solved examples, experiments on few-shot learning will be insightful. Dasgupta et al. (2022) reported that the models performed syllogistic reasoning better with few-shot learning, whether or not the content of the inferences were consistent with common-sense beliefs. In addition, we suppose that the performances can be different depending on the wording in instructions. Further research will contribute to improvement of instructions.

Second, we showed that the models performed worse in the cases in which humans have troubles because of some well-known human biases. It is left for future work to make more detailed comparisons between humans and neutral models, which is a promising research direction since the original BAROCO data on human reasoning ability is available.

Finally, it is interesting to consider extended forms of natural logic inferences other than basic syllogisms, including relational syllogism with transitive verbs and comparatives and those inferences with generalized quantifiers such as *most*. It is left for future work to examine whether the models show similar biases for such extended syllogism inferences.

Author Contributions

The three authors, Risako Ando, Takanobu Morishita, and Hirohiko Abe, made equal contributions as core authors.

Acknowledgements

We express our gratitude to the anonymous reviewers for their helpful feedback, which has contributed to the improvement of this work. We would like to express our thanks to Prof. Chizuru Shikishima for the discussion on earlier versions of the BAROCO inference tasks collection. This work is partially supported by JST, CREST grant number JPMJCR2114, MEXT-JSPS Kakenhi MKK279H and MKK477J.

References

Tom Brown, Benjamin Mann, Nick Ryder, Melanie Subbiah, Jared D Kaplan, Prafulla Dhariwal, Arvind Neelakantan, Pranav Shyam, Girish Sastry, Amanda Askell, et al. 2020. Language models are few-shot learners. *Advances in neural information processing systems*, 33:1877–1901.

Nick Chater and Mike Oaksford. 1999. The probability heuristics model of syllogistic reasoning. *Cognitive Psychology*, 38(2):191–258.

Ishita Dasgupta, Andrew K Lampinen, Stephanie CY Chan, Antonia Creswell, Dharshan Kumaran, James L McClelland, and Felix Hill. 2022. Language models show human-like content effects on reasoning. *arXiv preprint arXiv:2207.07051*.

Jacob Devlin, Ming-Wei Chang, Kenton Lee, and Kristina Toutanova. 2019. BERT: Pre-training of deep bidirectional transformers for language understanding. In *Proceedings of the 2019 Conference of the North American Chapter of the Association for Computational Linguistics: Human Language Technologies, Volume 1 (Long and Short Papers)*, pages 4171–4186, Minneapolis, Minnesota. Association for Computational Linguistics.

Jonathan St.B. T. Evans. 1989. *Bias in Human Reasoning: Causes and Consequences*. Lawrence Erlbaum Associates, Inc.

Jonathan St.B. T. Evans, Stephen E. Newstead, and Ruth M. J. Byrne. 1993. *Human Reasoning: The Psychology of Deduction*. Psychology Press.

Bart Geurts. 2003. Reasoning with quantifiers. *Cognition*, 86(3):223–251.

Vinod Goel, Christian Buchel, Chris Frith, and Raymond J Dolan. 2000. Dissociation of mechanisms underlying syllogistic reasoning. *Neuroimage*, 12(5):504–514.

Sangeet Khemlani and Philip N Johnson-Laird. 2012. Theories of the syllogism: A meta-analysis. *Psychological Bulletin*, 138(3):427.

Mike Lewis, Yinhan Liu, Naman Goyal, Marjan Ghazvininejad, Abdelrahman Mohamed, Omer Levy, Veselin Stoyanov, and Luke Zettlemoyer. 2020. BART: Denoising sequence-to-sequence pre-training for natural language generation, translation, and comprehension. In *Proceedings of the 58th Annual Meeting of the Association for Computational Linguistics*, pages 7871–7880, Online. Association for Computational Linguistics.

Yinhan Liu, Myle Ott, Naman Goyal, Jingfei Du, Mandar Joshi, Danqi Chen, Omer Levy, Mike Lewis, Luke Zettlemoyer, and Veselin Stoyanov. 2019. RoBERTa: A robustly optimized BERT pretraining approach. *arXiv preprint arXiv:1907.11692*.

Lawrence S. Moss. 2015. Natural logic. In Shalom Lappin and Chris Fox, editors, *The Handbook of Contemporary Semantic Theory*, 2 edition, pages 559–592. Wiley.

Stephen E. Newstead. 1989. Interpretational errors in syllogistic reasoning. *Journal of Memory and Language*, 28(1):78–91.

Stephen E Newstead, Paul Pollard, Jonathan St BT Evans, and Julie L Allen. 1992. The source of belief bias effects in syllogistic reasoning. *Cognition*, 45(3):257–284.

Kyle Richardson, Hai Hu, Lawrence Moss, and Ashish Sabharwal. 2020. Probing natural language inference models through semantic fragments. In *Proceedings of the AAAI Conference on Artificial Intelligence*, volume 34, pages 8713–8721.

Víctor Sánchez Valencia. 1991. *Studies on Natural Logic and Categorial Grammar*. Ph.D. thesis, University of Amsterdam.

Yuri Sato and Koji Mineshima. 2015. How diagrams can support syllogistic reasoning: an experimental study. *Journal of Logic, Language and Information*, 24:409–455.

Viktor Schlegel, Kamen Pavlov, and Ian Pratt-Hartmann. 2022. Can transformers reason in fragments of natural language? In *Proceedings of the 2022 Conference on Empirical Methods in Natural Language Processing*, pages 11184–11199, Abu Dhabi, United Arab Emirates. Association for Computational Linguistics.

Chizuru Shikishima, Juko Ando, Pierre Grialou, Ryo Takemura, and Mitsuhiro Okada. 2006. A behavioral genetic study of syllogism solving using linguistic and graphical representations: a preliminary report. In *Images and Reasoning*, pages 69–85. Keio University Press.

Chizuru Shikishima, Kai Hiraishi, Shinji Yamagata, Juko Ando, and Mitsuhiro Okada. 2015. Genetic factors of individual differences in decision making in economic behavior: A Japanese twin study using the Allais problem. *Frontiers in Psychology*, 6:1712.

Chizuru Shikishima, Kai Hiraishi, Shinji Yamagata, Yutaro Sugimoto, Ryo Takemura, Koken Ozaki, Mitsuhiro Okada, Tatsushi Toda, and Juko Ando. 2009. Is g an entity? a Japanese twin study using syllogisms and intelligence tests. *Intelligence*, 37(3):256–267.

Keith Stenning and Michiel Van Lambalgen. 2012. *Human Reasoning and Cognitive Science*. MIT Press.

Johan van Benthem. 1986. *Essays in Logical Semantics*. Reidel, Dordrecht.

Minna Cheves Wilkins. 1928. The effect of changed material on ability to do formal syllogistic reasoning. *Archives of Psychology*, 16:1–83.

Adina Williams, Nikita Nangia, and Samuel Bowman. 2018. A broad-coverage challenge corpus for sentence understanding through inference. In *Proceedings of the 2018 Conference of the North American Chapter of the Association for Computational Linguistics: Human Language Technologies, Volume 1 (Long Papers)*, pages 1112–1122, New Orleans, Louisiana. Association for Computational Linguistics.

Robert S Woodworth and Saul B. Sells. 1935. An atmosphere effect in formal syllogistic reasoning. *Journal of Experimental Psychology*, 18(4):451–460.

Hitomi Yanaka, Koji Mineshima, Daisuke Bekki, Kentaro Inui, Satoshi Sekine, Lasha Abzianidze, and Johan Bos. 2019. Can neural networks understand monotonicity reasoning? In *Proceedings of the 2019 ACL Workshop BlackboxNLP: Analyzing and Interpreting Neural Networks for NLP*, pages 31–40.

SpaceNLI: Evaluating the Consistency of Predicting Inferences in Space

Lasha Abzianidze **Joost Zwarts** **Yoad Winter**
Institute for Language Sciences, Utrecht University
Utrecht, the Netherlands
{l.abzianidze, j.zwarts, y.winter}@uu.nl

Abstract

While many natural language inference (NLI) datasets target certain semantic phenomena, e.g., negation, tense & aspect, monotonicity, and presupposition, to the best of our knowledge, there is no NLI dataset that involves diverse types of spatial expressions and reasoning. We fill this gap by semi-automatically creating an NLI dataset for spatial reasoning, called SpaceNLI.[1] The data samples are automatically generated from a curated set of reasoning patterns (see Figure 1), where the patterns are annotated with inference labels by experts. We test several SOTA NLI systems on SpaceNLI to gauge the complexity of the dataset and the system's capacity for spatial reasoning. Moreover, we introduce a *Pattern Accuracy* and argue that it is a more reliable and stricter measure than the accuracy for evaluating a system's performance on pattern-based generated data samples. Based on the evaluation results we find that the systems obtain moderate results on the spatial NLI problems but lack consistency per inference pattern. The results also reveal that non-projective spatial inferences (especially due to the "between" preposition) are the most challenging ones.

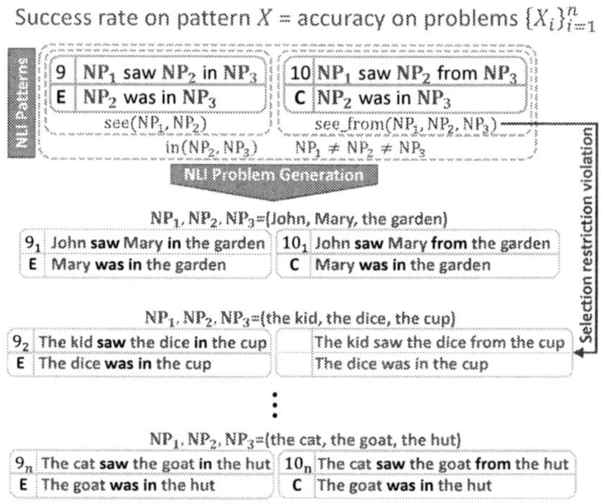

Figure 1: Sampling NLI problem from NLI patterns (with IDs 9 and 10, **E**ntailment and **C**ontradiction, respectively). The problems are generated by replacing NP placeholders with definite NPs that satisfy pattern-specific selection restrictions. A system's success rate on a pattern is defined as the accuracy on its corresponding NLI problems.

1 Introduction

Natural language inference (NLI) is a popular task that evaluates NLP systems on text reasoning skills. In the task, a system has to predict an inference relation from a premise text to a hypothesis sentence/phrase. Usually, the task is three- or two-way classification, depending on whether in the inference labels of *entailment*, *neutral*, and *contradiction*, the latter two are merged into *non-entailment*. The task is intended for evaluation of NLP systems on reasoning, however, the systems with competitive results on NLI benchmarks are often exploiting dataset biases (Tsuchiya 2018; Poliak et al. 2018; Gururangan et al. 2018; McCoy et al. 2019, *inter alia*) and their performance suffers from out-of-distribution NLI sample problems (Glockner et al., 2018).

To better evaluate the reasoning skills of NLI systems, a series of works have been (semi-)automatically or manually creating NLI datasets that specialize in certain semantic phenomena. While some of these datasets come with a training part, most of them are intended solely for evaluation. For example, several datasets have been dedicated to monotonicity reasoning (Yanaka et al., 2019b,a, 2020), negation was targeted by Hossain et al. (2020), the dataset by Kober et al. (2019) focuses on temporal and aspectual inferences, Jeretic et al. (2020) semi-automatically generated NLI problems for implicatures and presuppositions. There are also NLI datasets that cover several semantic phenomena, having a separate section for each of the phenomena (Cooper et al. 1996;

[1] https://github.com/kovvalsky/SpaceNLI

Proceedings of the 4th Natural Logic Meets Machine Learning Workshop (NALOMA23), pages 12–24
June 20, 2023.

Richardson et al. 2020, *inter alia*).

While spatial reasoning has been included in several multi-modal QA datasets (Antol et al., 2015; Suhr et al., 2017; Johnson et al., 2017; Hudson and Manning, 2019) and in a couple of text-based QA datasets (Weston et al., 2016; Mirzaee et al., 2021), to the best of our knowledge, no NLI dataset has specifically covered it.[2] This paper fills the gap by semi-automatically creating an NLI dataset for spatial inferences. First, we collected a diverse set of NLI problems inspired by the inference examples found in the literature on spatial semantics. Second, the NLI problems were manually converted into NLI patterns (see Figure 1), and finally, we automatically generated a large number of NLI problems from the patterns.

The paper makes two main contributions:

C1. SpaceNLI: the spatial NLI dataset with diverse types of spatial inferences; The inference labels of the generated problems are highly faithful (97%) to the labels of the corresponding original patterns.

C2. Pattern accuracy and its curve: they measure systems' performance on patterns and the consistency of predictions on samples from the same patterns.

The conducted experiments answer the following research questions:

Q1. How much spatial reasoning current SOTA NLI systems are capable of?

A1. We found out that the SOTA NLI systems have problems with fine-grained spatial inferences. Their performance drops at least by 24% compared to their results on common NLI datasets. Moreover, their consistency in predictions is sensitive to irrelevant lexical substitutions.

Q2. What types of spatial inference problems are easy or challenging for the SOTA NLI systems?

A2. The results showed that the non-projective spatial relations are most challenging for the models. This was mainly due to difficulty associated with "between" and its frequent occurrence in the evaluation dataset.

2 Spatial expressions and inferences

2.1 Types of spatial expressions

Spatial expressions consist of spatial prepositions and other expressions with spatial information (e.g., *far*, *the left of*, and *in front of*). They usually describe a relation between two entities, the *figure* and the *ground*. The site or path of the figure is the focus of the discussion and is characterized with respect to the ground. For example, in (9_1) and (10_1) from Figure 1, *Mary* is a figure and *garden* a ground. *John* is also a figure in the premise of (10_1).

Spatial expressions are roughly divided into *locative* and *directional* expressions, where locatives can be further classified into *projective* and *non-projective* (Herskovits, 1986). The locative expressions describe static, locative relations between the figure and the ground while directional ones describe a more *dynamic* relation involving a movement and/or path. An example with a directional preposition is *Cindi walked into the market.* The spatial expressions in Figure 1 are all locative except for *from*, which is directional. These locative expressions are non-projective since they require only the spatial location of the figure and the ground. In contrast, projective locatives additionally require further information from the ground in terms of a deictic frame of reference (i.e., an orientation structure). For example, the site of the house is not sufficient to interpret Mary's location in *Mary is behind the house*, it requires knowledge about the frame of reference of the house, in particular, what counts as a back side of the house.

2.2 Types of spatial inferences

We characterize spatial inferences depending on the type of spatial expressions licensing them. An inference might depend on several spatial expressions of a different type, which makes partitioning the inferences challenging, if not impossible. We define the following classes that represent a coarse-grained partition of spatial inferences. The classes will be later referred to in § 3.[3]

Argument orientation In spatial literature, an argument orientation entailment identifies which

[2] Even the FraCaS dataset (Cooper et al., 1996; MacCartney, 2009), which was curated by linguists and semanticists, doesn't cover spatial semantics within its nine sections.

[3] Licensing contradiction and neutral problems will be assumed from the perspective of a related entailment problem. For example, we assume that the neutral problem (16) in Table 1 is licensed in the same way as its related entailment (15). Put differently, one can see (16) as an adversary to (15) and assume that solving (15) requires competence comparable to the one required for solving (16).

ID	Class	Premise(s)	L	Hypothesis
15	Dir	John threw the ball into the box.	E	The ball went into the box.
16	Dir	John threw the ball at the box.	N	The ball went into the box.
31a	Dir	Los Angeles is in California. John came from California.	N	John came from Los Angeles.
38	NonP	John is in the garden. The garden is in the church.	E	John is in the church.
41	Dir	John drove through the tunnel.	E	John was in the tunnel.
47a	Dir	Cindi walked into the market.	E	Cindi was outside the market.
56c	Proj	The trash can is to the right of the tree from John.	C	The tree is to the right of the trash can from John.
70	Proj	Mary is between the tree and the house. The tree is behind the house.	E	Mary is behind the house.
80	NonP	The cat is between the house and the fence. The cat is between the fence and the tree.	C	The cat is between the house and the tree.
99*d	Proj	The bucket is above the bowl. The pencil is above the bowl.	N	The bucket is below the pencil.
96b	ArgO	Mary met John at the party.	N	Cindi was not at the party.
100	NonP	The house is far from the school.	E	The school is far from the house.
102a	ArgO	Mary has taken the cup out of the cabinet.	C	The cup is in the cabinet.
102f	ArgO	Mary has hidden the cup behind the cabinet.	E	The cup is not in the cabinet.

Table 1: Examples of the seed NLI problems annotated with spatial inference classes: **Dir**ectional, **Proj**ective, **Non-P**rojective, and **Arg**ument **O**rientation. Initial letters abbreviate the corresponding inference labels.

argument of the verb is the figure of the spatial expression. For instance, (9_1) in Figure 1 show that *Mary* is the figure of the locative PP *in the garden*. In its original interpretation, the argument orientation entailment is not restricted to spatial expressions of a particular type. Here, we restrict the class of argument orientation to the entailment problems (and their neutral and contradiction counterparts) that come close to resolving a PP attachment. For example, correctly resolving the PP attachment in (9_1) boils down to the hypothesis. The problems in this class contain a hypothesis with a copula and a predicative spatial PP, where the PP is contrasted to a tightly related PP in the premise(s). For more examples of the NLI problems in the argument orientation class, see Table 1.

Directional The directional class contains spatial inferences where directional spatial expressions play the key role. Examples of such inferences are given in Table 1. Some of these NLI problems pertain to a path-place relation: (47a) shows that *walking into* infers *being outside*;[4] (41) entails *being in the tunnel* from the premise that states that the driving path was through the tunnel. (31a) combines a part-whole relation with the movement path.

Projective This class contains inferences that hinge on a frame of reference introduced by projective spatial expressions. In principle, the frame of reference can introduce six directions that can be referred to using the expressions like *front, behind, left, right, above, below, under, on top of*, etc. (see the examples of NLI problems in Table 1). The NLI problems that contain *on top of* as only projective spatial expression, and when its projective interpretation is not crucial for the inference, are put in a different class.

Non-projective We classify a problem as having non-projective inference if the inference is driven only by non-projective spatial expressions. Therefore, an occurrence of non-projective spatial expressions in a problem is necessary but not sufficient for assigning the problem to this class, e.g., see directional problems (31a) and (41). NLI problems that depend on spatial expressions with the semantics of order and proximity are also in this class, see *between* (80) and *far* (100) in Table 1.

3 Dataset construction

3.1 Pattern construction

Patterns are labeled NLI problems with NPs replaced by variables as illustrated in Figure 1. The NLI patterns are obtained from the seed NLI problems. To collect the latter, we extracted the initial 56 problems from Zwarts and Winter (2000) and Nam (1995), where a majority of the problems were labeled as entailment due to obvious biases in the semantic literature towards licensing entail-

[4] Since moving along the path is related to the change of the location, sometimes spatial entailments interfere with tense and aspect.

Class (#patterns)	Spatial expression counts
Directional (95)	in (20), from (17), into (9), to (8), on (8), away from (7), towards (7), out of (4), back (3), through (3), across (2), at (2), outside (2), opposite (1), part of (1), by (1)
Argument orientation (67)	in (21), at (10), from (9), away from (4), out of (4), near (3), with (3), inside (3), on (2), under (2), through (1), opposite (1), towards (1), far from (1), on top of (1), behind (1)
Projective (70)	behind (16), between (11), in front of (10), below (6), above (6), under (6), on top of (5), front of (3), opposite (2), to the right of (2), on (2), to the left of (1)
Non-projective (48)	between (22), in (9), far from (5), close to (4), outside (3), on top of (2), on (2), opposite (1)

Table 2: The spatial expressions and their counts per entailment class in the SpaceNLI patterns

ment. To create a representative and challenging NLI dataset for machine learning, we applied several *revision phases* to the problems: introducing new problems that either cover new semantic aspects of spatial expression or serve as a perturbed version of an existing problem.

In the initial revision phase, four annotators divided the extracted problems and created slightly modified versions of them with an inference label different from the original.[5] This was motivated by the current trends in the literature on adversarial, stress, and debiased datasets (Naik et al. 2018; Ribeiro et al. 2020; Kaushik et al. 2020; Gardner et al. 2020, *inter alia*). For example, (16) is a perturbed example of (15). Where possible, NLI problems of a new type were also created using the similar spatial expressions found in the extracted problems.

To validate the resulting pool of NLI problems (in total 162), following (Zhang et al., 2017), they were labeled on a 5-point Likert scale by three annotators.[6] After collecting the 5-point annotations, for each annotator, we picked a mapping of 5-point to 3-point that maximizes the inter-annotator agreement (avg. Cohen's $\kappa = .71$). The problems without majority labels were discarded and 111 problems remained.

To better balance the inference labels and increase the coverage of spatial expressions, a second revision phase was carried out on the remaining problems. In several cases, problems with low annotator agreement were revised, e.g., changing the tense where it caused confusion or replacing a preposition with a weaker version (*at*$\mapsto$*near*). All the new and revised problems (in total 63) were validated based on three samples: each problem was manually converted into a pattern by replacing NPs with variables, and three random NLI samples per pattern were generated (see §3.2 for details), which were subsequently validated by three annotators.

Finally, a third revision phase was carried out on the remaining problems to additionally decrease the overall and spatial type-specific label imbalance. The collected problems (in total 160) were treated as a seed by converting them into NLI patterns to generate a large amount of sample NLI problems from them. To illustrate the coverage of spatial expressions in the collected patterns, Table 2 gives the complete list of spatial expressions for each entailment class.

3.2 Sample generation

We manually created NLI patterns from the initially collected NLI problems (§3.1) by replacing NPs with placeholders and specifying selection restrictions for them imposed by the verbs, spatial expressions, and gold inference labels (see Figure 1). The selection restrictions imposed by spatial expressions are subtle and can affect gold labels or the naturalness of sentences. For example, if the figure is much larger than the ground, it can make the sentence infelicitous: *the apple on the fridge* and *the apple near the fridge* are preferred to *the fridge under the apple* and *the fridge near the apple*. Inferences driven by proximity-related spatial expressions are sensitive to the size of the objects. For instance, based on our conducted validations, *Cindi is opposite to the cat* is more likely to be

[5]The annotators for the pattern construction consist of the authors of the paper, two linguist students, and one AI student. The guideline for creating inference problems can be found in the supplementary material.

[6]The question was to what extent the hypothesis sentence is true, given that the premises are true, with choices: *definitely false*, *most likely false*, *unknown*, *most likely true*, *definitely true*. We used two additional choices, *difficult* (unable to annotate due to the complex reasoning it requires) and *skip* (presence of an ungrammatical or nonsensical sentence). We used the brat annotation tool (Stenetorp et al., 2012) for labeling. The annotation guideline is included in the supplementary material.

neutral to *Cindi is far from the cat*, but *the school is opposite to the house* is more likely to contradict *the school is far from the house*.

To meet selection restrictions and allow relative diversity of NPs in the generated samples, we defined a mini world with a domain containing 171 entities corresponding to common and proper nouns. The entities are organized in a taxonomy with 20 subclasses covering general types of entities (e.g., person, animal, vehicle), the projections of an argument in certain argument structures (e.g., enter in X, be in X, throw X), compatibility with projective spatial expressions, and size categories (S for entities comparable to small objects like book and cat, M to persons, and L to vehicles). Binary and ternary relations are defined based on the set unions of the products of entity sets and subclasses.

To automatize the sampling of sound NLI problems from the patterns, we formatted the mini world in YAML and NLI patterns in XML. We implemented a procedure that samples problems from the patterns by filling in NP placeholders with definite NPs from the mini world and respecting the pattern-specific selection restrictions. For sanity checking, the procedure verifies that it can generate corresponding seed NLI problems for each pattern.

To measure how faithfully the inference labels are transferred from seed and pattern NLI problems to the corresponding NLI samples, we used sampled problems in the second phase of validation when validating new NLI problems (see §3.1). The results showed that 79% of samples were unanimously labeled with the original label. After filtering out patterns with a relatively low agreement, this ratio increased to 97% for the samples generated from the validated patterns.

The NLI problems sampled from the same pattern or related patterns are string-wise very close to each other, sometimes differing only in terms of occurrences of a single NP. Regardless of this similarity, we expect such problems to pose a challenge for NLI systems based on large language models (LLMs) as it has been shown that their predictions can be sensitive to a single-word substitution (Glockner et al., 2018; Gururangan et al., 2018). In addition to NPs, one could have allowed the replacement of other phrases in the NLI patterns, but this would have significantly complicated the definition of the mini world and generation of natural and sound NLI samples.

Property	E %	N %	C %	All % (#)	
Dir	39.6	35.4	25.0	30.0	(9600)
NonP	25.0	41.7	33.3	22.5	(7200)
Proj	29.4	26.5	44.1	21.2	(6800)
ArgO	47.6	28.6	23.8	26.2	(8400)
+ neg	48.0	28.0	24.0	15.6	(5000)
1prem	41.8	26.5	31.6	61.3	(19600)
2prem	25.0	42.9	32.1	35.0	(11200)
3prem	50.0	50.0	0.0	3.8	(1200)
All	36.2	33.1	30.6	100.0	(32000)

Table 3: Statistics of several properties of the sampled NLI dataset. The statistics also apply to the collection of NLI patterns as the samples are evenly distributed over the patterns. The properties consist of the spatial inference types, whether including negation, and the number of premises.

LLM-based NLI models	Training data	SNLI + MNLI	SpaceNLI Acc	SpaceNLI $PA_{0.95}$	SpaceNLI $PA_{1.0}$
DeBERTaV3-L#1 Joelzhang/deberta-v3...	SMFA	**91.8**	59.6	47.5	**37.5**
ALBERT-XXLv2 ynie/albert-xxlarge-v2...	SMFA	90.8	57.8	**48.1**	36.2
DeBERTa-L He et al. (2021)	M	90.7	54.1	42.5	36.2
RoBERTa-L Nie et al. (2020)	SMFA	90.6	55.6	40.0	31.9
BART-L ynie/bart-large-snli_mnli...	SMFA	90.4	55.4	39.4	29.4
DeBERTaV3-L#2 Laurer et al. (2022)	MFALW	90.3	**66.5**	44.4	33.8
XLNet-L-cased Nie et al. (2020)	SMFA	90.3	55.8	42.5	30.0

Table 4: Performance of SOTA NLI systems on SpaceNLI. SNLI+MNLI shows the average score on these datasets. Training data names are denoted with the initial letters: **S**NLI, **M**NLI, **A**NLI, **F**ever-NLI, **W**ANLI, and **L**ingNLI. The best system per problem accuracy on SpaceNLI, DeBERTaV3-L_{MFALW} (with $\Delta \geq 6.9\%$), doesn't turn out to be the best at the consistency threshold ≥ 0.95. Table 5 in Appendix A represents an extended version of the table with more threshold points.

4 Experiments

4.1 Sample dataset

We uniformly generated a spatial dataset of 32,000 NLI samples from 160 NLI patterns, i.e., 200 samples per pattern. We used the mini world as described in § 3.2. The dataset statistics are given in Table 3. The inference labels are relatively balanced: each label being represented by at least 30% of the problems. Each spatial inference type counts at least 20% of the overall problems and 23% of label-specific problems. In contrast to the common biases in NLI datasets, a majority of the problems with negation are labeled as entailment, not contradiction. This is due to perturbed problems introduced in the revision phases (§ 3.1). Around 39% of problems have multiple premises, where three-premised problems occur only in the directional problems, the argument orientation problems contain only single-premised problems, and most of the multi-premised problems are in the non-projective problems. We refer to the generated dataset as SpaceNLI and use it in subsequent experiments.[7]

4.2 Evaluating SOTA NLI systems

4.2.1 Standard accuracy

We selected NLI models that have results comparable to the state of the art in NLI and evaluate them on SpaceNLI. The models were chosen based on their availability, tractable size, and high average accuracy (> 90%) on the SNLI (Bowman et al., 2015) and MNLI (Williams et al., 2018) datasets (see Table 4). The models are based on various large language models (LLMs) like DeBERTaV3 (He et al., 2023), BART (Lewis et al., 2020), ALBERT (Lan et al., 2020), XLNet (Yang et al., 2020), etc. (see Table 4). The LLMs are fine-tuned on several NLI train datasets: SNLI, MNLI, FEVER-NLI (Nie et al., 2019), ANLI (Nie et al., 2020), LingNLI (Parrish et al., 2021), WANLI (Liu et al., 2022). We use the models from the HuggungFace model hub[8] and provide them with the corresponding hub names in Table 4.

The results in Table 4 show that DeBERTaV3-L#2 trained on a large collection of training datasets (885K problems in total) generalizes best on the spatial reasoning (66.5%), achieving a substantial improvement (≥ 6.9%) over the other models.[9]

4.2.2 Consistency & pattern accuracy

To evaluate the models on the consistency of their predictions for NLI problems from the same pattern, we define the pattern accuracy (PA) score and its curve. The PA curve records the PA score of a model for each consistency threshold. Informally, the PA score with a consistency threshold t is a ratio of NLI patterns for which model gets at least t portion of the samples generated from them. For example, the PA of 50% with a threshold 90% means that there are a half of the NLI patterns such that for each pattern a model is able to correctly classify at least 90% of its sample problems. The formal definition of the PA with a threshold t is:

$$PA_t(\hat{Y}, \mathbf{y}) = \frac{1}{N} \sum_{i=1}^{N} \left[\frac{\sum_{k=1}^{M_i} \delta(\hat{y}_k^i = y^i)}{M_i} \geq t \right]$$

where $\hat{Y} = (\hat{y}_k^i)_{1 \leq i \leq N, 1 \leq k \leq M_i}$ are predictions for k^{th} sample of ith pattern, N is the number of patterns, M_i is the number of samples for i^{th} pattern, $\mathbf{y} = (y^i)_{1 \leq i \leq N}$ gold labels of ith pattern, and δ is the Kronecker delta.

While DeBERTaV3-L#2 gets the best score on the SpaceNLI problems, based on the PA scores in Table 4, it shows high consistency ($PA_{0.95}$ or $PA_{1.0}$) in fewer NLI patterns than the other two competing models, DeBERTaV3-L#1 and ALBERT-XXLv2. PA curves of the NLI models provide a closer look at this contrast (see Figure 2). While the curve of DeBERTaV3-L#2 outperforms other models by a margin, it is noteworthy that it does this by classifying sample problems of the patterns which it can hardly solve half of the

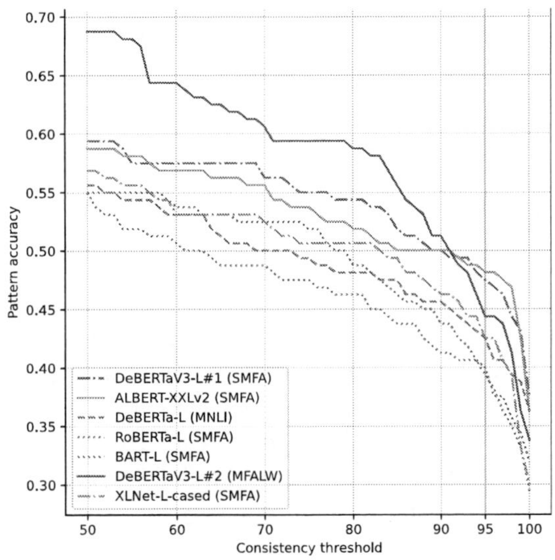

Figure 2: Pattern accuracy curves of the NLI models from Table 4. The first half, which corresponds to the scores allowing solving less than half of the samples per pattern, is omitted (see Figure 6 in Appendix A for the complete curves).

[7]We make the collection of the patterns, the generation code, and the sample dataset publicly available upon the acceptance of the paper.

[8]`https://huggingface.co/models`

[9]The second best, DeBERTaV3-L#1, is based on the same LLM fine-tuned on a different combination of NLI datasets. Note that Laurer et al. (2022) deliberately removed SNLI from the training set as it negatively affected the accuracy of the model in their experiments.

time (this is visible in the complete curves of Figure 6 in Appendix A). It drastically decreases after 95% of consistency while ALBERT-XXLv2 and DeBERTAV2-L#1 maintain very high consistency for $> 47\%$ of NLI patterns. This demonstrates that a high-performing model is not necessarily the most consistent across patterns.

RoBERTa-L and BART-L obtain similar accuracy scores, but RoBERTa-L is more consistent in more NLI patterns than BART-L while the latter gets slightly more NLI problems for inconsistently predicted patterns. The complete curves of Figure 6 in Appendix A show how the curves swap places after the consistency threshold of 50. This shows that the standard accuracy (i.e., based on NLI problem samples) can blur the fine distinction in consistency between the models.

The dispersion of the curves at the lowest end of the consistency threshold is twice larger than at the highest end. This shows that the model predictions more diverge in coverage of patterns than in consistency per pattern. In other words, the contrast confirms the sensitivity of the models towards the inference-preserving word substitutions.

4.2.3 Few-shot learning experiments

We measured the difficulty of the SpaceNLI problems in terms of few-shot learning experiments. We used 100 samples per pattern as a test set while other 100 samples per pattern were used for drawing a few samples for each pattern. In this way, the patterns are fully shared between the training and test sets, but no sample NLI problem is in both sets. For each number of shots, we carried out the sample drawing process three times. We used two NLI models: a high performing NLI model RoBERTa-L_{SMFA} from Nie et al. (2020) and a *vanilla* NLI model based on the large RoBERTa pretrained language model (Liu et al., 2019). The results of the few-shot experiments are in Figure 3.

Finetuning RoBERTa-L_{SMFA} on a single sample of each pattern increases the sample-based accuracy on the test set by 14%. Each additional sample further boosts the model's accuracy. The almost perfect accuracy (>99%) is reached when 20 samples per pattern are seen during the finetuning. The results show that the lexical variability poses a challenge to the high-performing NLI model as it needs to be finetuned on at least five samples for every pattern of the test set to achieve a high score.

The challenge coming from the lexical variability and the SpaceNLI patterns is further emphasized by the relatively low results of RoBERTa Large. Even after being finetuned on the 20 samples of each NLI pattern, the model is still far from the high performance on unseen samples (but seen patterns). The relatively low results can be also partially attributed to the low ratio between the number of training samples and the large number of the model's trainable parameters.

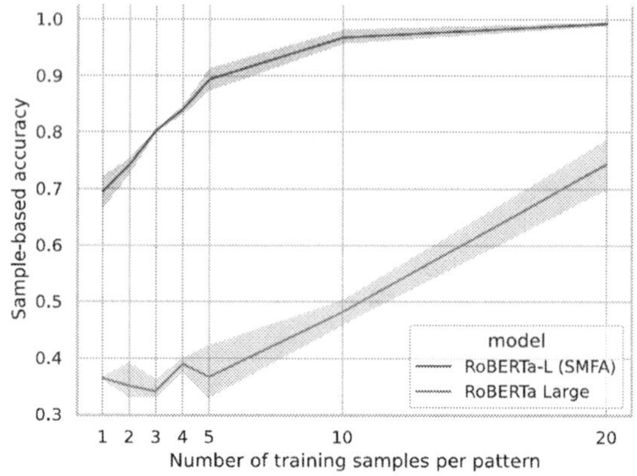

Figure 3: Average of three runs for each few-shot finetuning experiment. RoBERTa-L (SMFA, Nie et al. 2020) is already finetuned on several large NLI datasets while RoBERTa Large (Liu et al., 2019) is a pretrained language model without any previous training on NLI.

5 Analysis

To find out what type of inferences the models find challenging, we analyze the models' performance per inference type. Figure 5 shows the sample- and pattern-based accuracy scores of the models per spatial inference types as defined in §2.2. The model ranking based on the sample accuracy varies across the inference types. For instance, the best model, DeBERTaV3-L#2, remains at the top of the rankings for all inference types with quite a margin except for the projective type. On average, non-projective spatial inferences are the most challenging for the models. The easiest of the types is argument orientation, the type that is closest to the PP attachment task. For the other inference types, projective inferences are harder than directional ones. The apparent distinction in the scores between the inference types is also preserved for the $PA_{0.95}$ score (shown with the dark bars in Figure 5). The fine-grained analysis additionally shows that the best model, DeBERTaV3-L#2, suffers least in terms of consistency on the projective inferences while its performance on this inference type is not among the best.

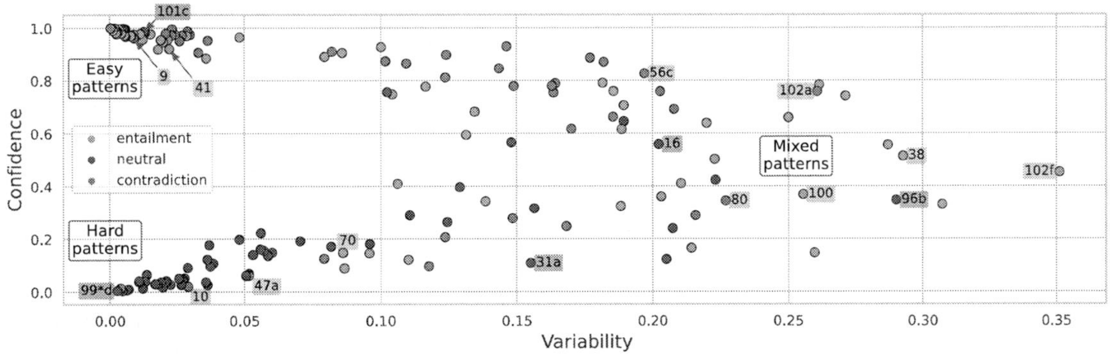

Figure 4: Prediction cartography of `RoBERTa-large` from (Nie et al., 2020). NLI patterns are characterized with *confidence* and *variability*: the mean and the standard deviation of probabilities assigned by the model to the true labels of the sample NLI problems. IDs mark NLI patterns from Figure 1 and Table 1.

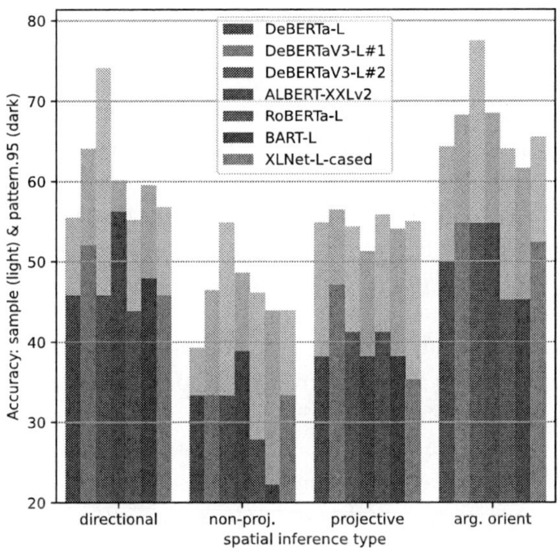

Figure 5: Sample-based (in light shades) and $PA_{0.95}$ (in dark shades) accuracy scores of the models per spatial inference type.

Based on the results in Figure 5, the non-projective NLI patterns and samples are the most challenging for the SOTA models. When looking closer at the set of non-projective problems, it turns out that it contains a high number of problems (46%) with the spatial expression "between" (as shown in Table 2), and these problems are specially challenging due to complex semantics of "between". The average accuracy of the models on such NLI samples is 41.6%. This is lower than the average sample-based accuracy (46.1%) on entire SpaceNLI and much lower than the average sample-based accuracy (54.1%) on the other part of the non-projective samples.

We further zoom in on the NLI patterns and measure a model's probabilistic predictions for the patterns. Namely, following Swayamdipta et al. (2020), we measure a model's confidence and variability. Originally the dataset cartography (Swayamdipta et al., 2020) was used to analyze the training dynamics of a model across the epochs and identify training samples that are easy or difficult for learning. In contrast, we use dataset cartography for analyzing evaluation dynamics across patterns and identifying easy and hard ones.[10]

Figure 4 illustrates the pattern-based evaluation dynamics of RoBERTa-L (Nie et al., 2020), an average model based on the evaluations. For instance, NLI pattern (102f) happens to have one of the most variable samples according to the model predictions: the mean and the standard deviation of the probabilities the model assigns to the entailment class of the samples of (102f) are 0.45 and 0.35, respectively.

(102f) NP_1 has hidden NP_2 behind NP_3.
`entailment` NP_2 is not in NP_3.

The evaluation cartography shows that the predictions vary mostly for entailment patterns (in green). Most of the hard patterns are neutral ones (in blue) and vice versa. Contradiction patterns (in red) tend to be easy with some variability.

6 Related work

Several works have automatically sampled NLI problems from curated patterns/templates. Jeretic et al. (2020) generated the implicature and presupposition diagnostic dataset IMPPRES from predefined templates. McCoy et al. (2019) constructed

[10] Put differently, iterative classification of the same training sample across epochs, is replaced with the classification of the same NLI pattern based on its samples.

the HANS dataset by designing templates of NLI problems that support or refute certain inference heuristics, which were later used to generate NLI problems. Richardson et al. (2020) used the template language from Salvatore et al. (2019) to produce NLI problems involving negation, Boolean connectives, quantifiers, cardinals, conditionals, and comparatives. These works all use restricted vocabulary while generating samples from the patterns.

With its pattern-based construction and restricted vocabulary, SpaceNLI comes close to the IMPPRES (Jeretic et al., 2020) and HANS (McCoy et al., 2019) datasets. Unlike these datasets, SpaceNLI involves multiple-premised problems and puts more emphasis on satisfying selection restrictions to prevent nonsensical sentences.

Based on the nature of NLI problems, SpaceNLI resembles FraCaS (Cooper et al., 1996) as both contain inference problems often found in textbooks on formal semantics. Unlike FraCaS, the inference labels of patterns in SpaceNLI are quite balanced and the number of spatial NLI patterns is twice the size of the largest section in FraCaS.

There have been attempts to identify semantic phenomena in existing NLI datasets, including aspects of spatial reasoning. By looking up certain keywords, Kim et al. (2019) automatically detect NLI problems in MultiNLI (Williams et al., 2018) that might contain spatial expressions. They create a mutated sample from the original NLI problem by negating the sentence with the potential spatial expression. Joshi et al. (2020) annotate MultiNLI problems based on the semantic aspects required by the inference label. Their taxonomic categories include the spatial subcategory, grouped with the relational, temporal, causal, and co-reference subcategories.

The problems in SpaceNLI are substantially diverse from a semantic perspective than the MultiNLI problems that were identified by Kim et al. (2019) and Joshi et al. (2020). The MultiNLI dataset is crowd-elicited and doesn't have problems with sufficient depth in spatial reasoning.

7 Conclusion

To the best of our knowledge, we have created the first spatial inference dataset that involves diverse spatial inference types. The structure and the evaluation protocol are unique as we focus on performance on the NLI patterns and consistency across the samples in the pattern, instead of focusing on mere quantitative accuracy based on the NLI problems/samples. The evaluation protocol tests models whether they can consistently recognize inference patterns while generalizing over *irrelevant* lexical substitutions. The more consistent a model is in its predictions, the less unexpected its behavior becomes.

The SOTA NLI models show moderate generalization capacity on spatial problems. While the top-performing model gets the highest overall accuracy, it is ranked third when it comes to the consistency of predictions inside the patterns: predicting at least 95% of the samples per pattern.

The introduced pattern accuracy (PA) curves provide a more fine-grained distinction between the models: the models with comparable standard accuracy scores might substantially differ in the consistency of their predictions. Overall the performance of models drops ca. 10% when raising the consistency threshold to 95%. This illustrates that the predictions of the SOTA models are sensitive to lexical replacements that have no effect on the semantics of the inference.

The evaluation results revealed that the most challenging inference type is associated with non-projective locatives mainly due to the complex semantics of "between" while the argument orientation type is the easiest. The latter is somewhat expected as the problems in the argument orientation type are close to the task of PP attachment which LLMs are expected to be good at.

Acknowledgments

This work was funded by the European Research Council (ERC) under the European Union's Horizon 2020 research and innovation programme (grant agreement No 742204). We would like to acknowledge the help from three student assistants with the data annotation and thank the anonymous reviewers for their helpful comments.

References

Stanislaw Antol, Aishwarya Agrawal, Jiasen Lu, Margaret Mitchell, Dhruv Batra, C. Lawrence Zitnick, and Devi Parikh. 2015. VQA: Visual Question Answering. In *International Conference on Computer Vision (ICCV)*.

Samuel R. Bowman, Gabor Angeli, Christopher Potts, and Christopher D. Manning. 2015. A large annotated corpus for learning natural language inference.

In *Proceedings of the 2015 Conference on Empirical Methods in Natural Language Processing*, pages 632–642, Lisbon, Portugal. Association for Computational Linguistics.

Robin Cooper, Dick Crouch, Jan Van Eijck, Chris Fox, Josef Van Genabith, Jan Jaspars, Hans Kamp, David Milward, Manfred Pinkal, Massimo Poesio, Steve Pulman, Ted Briscoe, Holger Maier, and Karsten Konrad. 1996. *FraCaS: A Framework for Computational Semantics*. Deliverable D16.

Matt Gardner, Yoav Artzi, Victoria Basmov, Jonathan Berant, Ben Bogin, Sihao Chen, Pradeep Dasigi, Dheeru Dua, Yanai Elazar, Ananth Gottumukkala, Nitish Gupta, Hannaneh Hajishirzi, Gabriel Ilharco, Daniel Khashabi, Kevin Lin, Jiangming Liu, Nelson F. Liu, Phoebe Mulcaire, Qiang Ning, Sameer Singh, Noah A. Smith, Sanjay Subramanian, Reut Tsarfaty, Eric Wallace, Ally Zhang, and Ben Zhou. 2020. Evaluating models' local decision boundaries via contrast sets. In *Findings of the Association for Computational Linguistics: EMNLP 2020*, pages 1307–1323, Online. Association for Computational Linguistics.

Max Glockner, Vered Shwartz, and Yoav Goldberg. 2018. Breaking NLI systems with sentences that require simple lexical inferences. In *Proceedings of the 56th Annual Meeting of the Association for Computational Linguistics (Volume 2: Short Papers)*, pages 650–655, Melbourne, Australia. Association for Computational Linguistics.

Suchin Gururangan, Swabha Swayamdipta, Omer Levy, Roy Schwartz, Samuel Bowman, and Noah A. Smith. 2018. Annotation artifacts in natural language inference data. In *Proceedings of the 2018 Conference of the North American Chapter of the Association for Computational Linguistics: Human Language Technologies, Volume 2 (Short Papers)*, pages 107–112, New Orleans, Louisiana. Association for Computational Linguistics.

Pengcheng He, Jianfeng Gao, and Weizhu Chen. 2023. DeBERTav3: Improving deBERTa using ELECTRA-style pre-training with gradient-disentangled embedding sharing. In *The Eleventh International Conference on Learning Representations*.

Pengcheng He, Xiaodong Liu, Jianfeng Gao, and Weizhu Chen. 2021. Deberta: Decoding-enhanced bert with disentangled attention.

Annette Herskovits. 1986. *Language and Spatial Cognition: an interdisciplinary study of the prepositions in English*. Studies in Natural Language Processing. Cambridge University Press, London.

Md Mosharaf Hossain, Venelin Kovatchev, Pranoy Dutta, Tiffany Kao, Elizabeth Wei, and Eduardo Blanco. 2020. An analysis of natural language inference benchmarks through the lens of negation. In *Proceedings of the 2020 Conference on Empirical Methods in Natural Language Processing (EMNLP)*, pages 9106–9118, Online. Association for Computational Linguistics.

Drew A Hudson and Christopher D Manning. 2019. Gqa: A new dataset for real-world visual reasoning and compositional question answering. In *Proceedings of the IEEE Conference on Computer Vision and Pattern Recognition*, pages 6700–6709.

Paloma Jeretic, Alex Warstadt, Suvrat Bhooshan, and Adina Williams. 2020. Are natural language inference models IMPPRESsive? Learning IMPlicature and PRESupposition. In *Proceedings of the 58th Annual Meeting of the Association for Computational Linguistics*, pages 8690–8705, Online. Association for Computational Linguistics.

Justin Johnson, Bharath Hariharan, Laurens van der Maaten, Li Fei-Fei, C. Lawrence Zitnick, and Ross Girshick. 2017. Clevr: A diagnostic dataset for compositional language and elementary visual reasoning. In *Proceedings of the IEEE Conference on Computer Vision and Pattern Recognition (CVPR)*.

Pratik Joshi, Somak Aditya, Aalok Sathe, and Monojit Choudhury. 2020. TaxiNLI: Taking a ride up the NLU hill. In *Proceedings of the 24th Conference on Computational Natural Language Learning*, pages 41–55, Online. Association for Computational Linguistics.

Divyansh Kaushik, Eduard Hovy, and Zachary C Lipton. 2020. Learning the difference that makes a difference with counterfactually augmented data. *International Conference on Learning Representations (ICLR)*.

Najoung Kim, Roma Patel, Adam Poliak, Patrick Xia, Alex Wang, Tom McCoy, Ian Tenney, Alexis Ross, Tal Linzen, Benjamin Van Durme, Samuel R. Bowman, and Ellie Pavlick. 2019. Probing what different NLP tasks teach machines about function word comprehension. In *Proceedings of the Eighth Joint Conference on Lexical and Computational Semantics (*SEM 2019)*, pages 235–249, Minneapolis, Minnesota. Association for Computational Linguistics.

Thomas Kober, Sander Bijl de Vroe, and Mark Steedman. 2019. Temporal and aspectual entailment. In *Proceedings of the 13th International Conference on Computational Semantics - Long Papers*, pages 103–119, Gothenburg, Sweden. Association for Computational Linguistics.

Zhenzhong Lan, Mingda Chen, Sebastian Goodman, Kevin Gimpel, Piyush Sharma, and Radu Soricut. 2020. Albert: A lite bert for self-supervised learning of language representations.

Moritz Laurer, Wouter van Atteveldt, Andreu Casas, and Kasper Welbers. 2022. Less Annotating, More Classifying – Addressing the Data Scarcity Issue of Supervised Machine Learning with Deep Transfer Learning and BERT-NLI.

Mike Lewis, Yinhan Liu, Naman Goyal, Marjan Ghazvininejad, Abdelrahman Mohamed, Omer Levy, Veselin Stoyanov, and Luke Zettlemoyer. 2020. BART: Denoising sequence-to-sequence pre-training for natural language generation, translation, and comprehension. In *Proceedings of the 58th Annual Meeting of the Association for Computational Linguistics*, pages 7871–7880, Online. Association for Computational Linguistics.

Alisa Liu, Swabha Swayamdipta, Noah A. Smith, and Yejin Choi. 2022. WANLI: Worker and AI collaboration for natural language inference dataset creation. In *Findings of the Association for Computational Linguistics: EMNLP 2022*, pages 6826–6847, Abu Dhabi, United Arab Emirates. Association for Computational Linguistics.

Yinhan Liu, Myle Ott, Naman Goyal, Jingfei Du, Mandar Joshi, Danqi Chen, Omer Levy, Mike Lewis, Luke Zettlemoyer, and Veselin Stoyanov. 2019. Roberta: A robustly optimized bert pretraining approach. Cite arxiv:1907.11692.

Bill MacCartney. 2009. *Natural language inference*. Phd thesis, Stanford University.

Tom McCoy, Ellie Pavlick, and Tal Linzen. 2019. Right for the wrong reasons: Diagnosing syntactic heuristics in natural language inference. In *Proceedings of the 57th Annual Meeting of the Association for Computational Linguistics*, pages 3428–3448, Florence, Italy. Association for Computational Linguistics.

Roshanak Mirzaee, Hossein Rajaby Faghihi, Qiang Ning, and Parisa Kordjamshidi. 2021. SPARTQA: A textual question answering benchmark for spatial reasoning. In *Proceedings of the 2021 Conference of the North American Chapter of the Association for Computational Linguistics: Human Language Technologies*, pages 4582–4598, Online. Association for Computational Linguistics.

Aakanksha Naik, Abhilasha Ravichander, Norman Sadeh, Carolyn Rose, and Graham Neubig. 2018. Stress test evaluation for natural language inference. In *Proceedings of the 27th International Conference on Computational Linguistics*, pages 2340–2353, Santa Fe, New Mexico, USA. Association for Computational Linguistics.

Seungho Nam. 1995. *The Semantics of Locative Prepositional Phrases in English.* Phd thesis, University of California, Los Angeles.

Yixin Nie, Haonan Chen, and Mohit Bansal. 2019. Combining fact extraction and verification with neural semantic matching networks. In *Association for the Advancement of Artificial Intelligence (AAAI)*.

Yixin Nie, Adina Williams, Emily Dinan, Mohit Bansal, Jason Weston, and Douwe Kiela. 2020. Adversarial NLI: A new benchmark for natural language understanding. In *Proceedings of the 58th Annual Meeting of the Association for Computational Linguistics*, pages 4885–4901, Online. Association for Computational Linguistics.

Alicia Parrish, William Huang, Omar Agha, Soo-Hwan Lee, Nikita Nangia, Alexia Warstadt, Karmanya Aggarwal, Emily Allaway, Tal Linzen, and Samuel R. Bowman. 2021. Does putting a linguist in the loop improve NLU data collection? In *Findings of the Association for Computational Linguistics: EMNLP 2021*, pages 4886–4901, Punta Cana, Dominican Republic. Association for Computational Linguistics.

Adam Poliak, Jason Naradowsky, Aparajita Haldar, Rachel Rudinger, and Benjamin Van Durme. 2018. Hypothesis only baselines in natural language inference. In *Proceedings of the Seventh Joint Conference on Lexical and Computational Semantics*, pages 180–191, New Orleans, Louisiana. Association for Computational Linguistics.

Marco Tulio Ribeiro, Tongshuang Wu, Carlos Guestrin, and Sameer Singh. 2020. Beyond accuracy: Behavioral testing of NLP models with CheckList. In *Proceedings of the 58th Annual Meeting of the Association for Computational Linguistics*, pages 4902–4912, Online. Association for Computational Linguistics.

Kyle Richardson, Hai Hu, Lawrence S. Moss, and Ashish Sabharwal. 2020. Probing natural language inference models through semantic fragments. In *The Thirty-Fourth AAAI Conference on Artificial Intelligence, AAAI 2020, The Thirty-Second Innovative Applications of Artificial Intelligence Conference, IAAI 2020, The Tenth AAAI Symposium on Educational Advances in Artificial Intelligence, EAAI 2020, New York, NY, USA, February 7-12, 2020*, pages 8713–8721. AAAI Press.

Felipe Salvatore, Marcelo Finger, and Roberto Hirata Jr. 2019. A logical-based corpus for cross-lingual evaluation. In *Proceedings of the 2nd Workshop on Deep Learning Approaches for Low-Resource NLP (DeepLo 2019)*, pages 22–30, Hong Kong, China. Association for Computational Linguistics.

Pontus Stenetorp, Sampo Pyysalo, Goran Topić, Tomoko Ohta, Sophia Ananiadou, and Jun'ichi Tsujii. 2012. brat: a web-based tool for NLP-assisted text annotation. In *Proceedings of the Demonstrations at the 13th Conference of the European Chapter of the Association for Computational Linguistics*, pages 102–107, Avignon, France. Association for Computational Linguistics.

Alane Suhr, Mike Lewis, James Yeh, and Yoav Artzi. 2017. A corpus of natural language for visual reasoning. In *Proceedings of the 55th Annual Meeting of the Association for Computational Linguistics (Volume 2: Short Papers)*, pages 217–223, Vancouver, Canada. Association for Computational Linguistics.

Swabha Swayamdipta, Roy Schwartz, Nicholas Lourie, Yizhong Wang, Hannaneh Hajishirzi, Noah A. Smith, and Yejin Choi. 2020. Dataset cartography: Mapping and diagnosing datasets with training dynamics. In *Proceedings of the 2020 Conference on Empirical Methods in Natural Language Processing (EMNLP)*, pages 9275–9293, Online. Association for Computational Linguistics.

Masatoshi Tsuchiya. 2018. Performance impact caused by hidden bias of training data for recognizing textual entailment. In *Proceedings of the Eleventh International Conference on Language Resources and Evaluation (LREC 2018)*, Miyazaki, Japan. European Language Resources Association (ELRA).

Jason Weston, Antoine Bordes, Sumit Chopra, and Tomás Mikolov. 2016. Towards ai-complete question answering: A set of prerequisite toy tasks. In *4th International Conference on Learning Representations, ICLR 2016, San Juan, Puerto Rico, May 2-4, 2016, Conference Track Proceedings*.

Adina Williams, Nikita Nangia, and Samuel Bowman. 2018. A broad-coverage challenge corpus for sentence understanding through inference. In *Proceedings of the 2018 Conference of the North American Chapter of the Association for Computational Linguistics: Human Language Technologies, Volume 1 (Long Papers)*, pages 1112–1122. Association for Computational Linguistics.

Hitomi Yanaka, Koji Mineshima, Daisuke Bekki, and Kentaro Inui. 2020. Do neural models learn systematicity of monotonicity inference in natural language? In *Proceedings of the 58th Annual Meeting of the Association for Computational Linguistics*, pages 6105–6117, Online. Association for Computational Linguistics.

Hitomi Yanaka, Koji Mineshima, Daisuke Bekki, Kentaro Inui, Satoshi Sekine, Lasha Abzianidze, and Johan Bos. 2019a. Can neural networks understand monotonicity reasoning? In *Proceedings of the 2019 ACL Workshop BlackboxNLP: Analyzing and Interpreting Neural Networks for NLP*, pages 31–40, Florence, Italy. Association for Computational Linguistics.

Hitomi Yanaka, Koji Mineshima, Daisuke Bekki, Kentaro Inui, Satoshi Sekine, Lasha Abzianidze, and Johan Bos. 2019b. HELP: A dataset for identifying shortcomings of neural models in monotonicity reasoning. In *Proceedings of the Eighth Joint Conference on Lexical and Computational Semantics (*SEM 2019)*, pages 250–255, Minneapolis, Minnesota. Association for Computational Linguistics.

Zhilin Yang, Zihang Dai, Yiming Yang, Jaime Carbonell, Ruslan Salakhutdinov, and Quoc V. Le. 2020. Xlnet: Generalized autoregressive pretraining for language understanding.

Sheng Zhang, Rachel Rudinger, Kevin Duh, and Benjamin Van Durme. 2017. Ordinal common-sense inference. *Transactions of the Association for Computational Linguistics*, 5:379–395.

Joost Zwarts and Yoad Winter. 2000. Vector space semantics: A model-theoretic analysis of locative prepositions. *Journal of logic, language and information*, 9:169–211.

A Results

LLM-based NLI models (train data) model names from Huggingface hub	SNLI	M_m	M_{mm}	S+M	SpaceNLI (accuracy & ≥ consistency score) Acc	≥ 0.5	≥ 0.67	≥ 0.9	≥ 0.95	$= 1.0$
DeBERTaV3-L#1 (SMFA) Joelzhang/deberta-v3-large-snli_mnli_fever_anli...	**92.9**	**91.4**	**91.2**	**91.8**	59.6	59.4	57.5	50.0	47.5	**37.5**
ALBERT-XXLv2 (SMFA) ynie/albert-xxlarge-v2-snli_mnli_fever_anli_R1_...	91.9	90.2	90.2	90.8	57.8	58.8	56.2	50.0	**48.1**	36.2
DeBERTa-L (MNLI) (He et al., 2021) microsoft/deberta-large-mnli	89.6	91.3	91.1	90.7	54.1	55.6	50.6	45.6	42.5	36.2
RoBERTa-L (SMFA) (Nie et al., 2020) ynie/roberta-large-snli_mnli_fever_anli_R1_R2_R...	91.8	89.9	90.0	90.6	55.6	55.0	52.5	43.8	40.0	31.9
BART-L (SMFA) ynie/bart-large-snli_mnli_fever_anli_R1_R2_R3-nli	92.0	89.4	89.6	90.4	55.4	55.0	48.8	41.2	39.4	29.4
DeBERTaV3-L#2 (MFALW) (Laurer et al., 2022) MoritzLaurer/DeBERTa-v3-large-mnli-fever-anli-l...	89.0	91.2	90.8	90.3	**66.5**	**68.8**	**61.9**	**51.2**	44.4	33.8
XLNet-L-cased (SMFA) (Nie et al., 2020) ynie/xlnet-large-cased-snli_mnli_fever_anli_R1_...	91.7	89.8	89.5	90.3	55.8	56.9	53.1	46.2	42.5	30.0

Table 5: Performance of NLI models on SpaceNLI and common NLI benchmarks: SNLI-test, MNLI-val-matched, and MNLI-val-mismatched. S+M shows the average of the three accuracy scores. Training data names are denoted with the initial letters: **S**NLI, **M**NLI, **A**NLI, **F**ever-NLI, **W**ANLI, and **L**ingNLI. The best model per problem accuracy on SpaceNLI, DeBERTaV3-L_{MFALW} (with $\Delta \geq 6.9\%$), doesn't turn out to be the best at the consistency threshold ≥ 0.95.

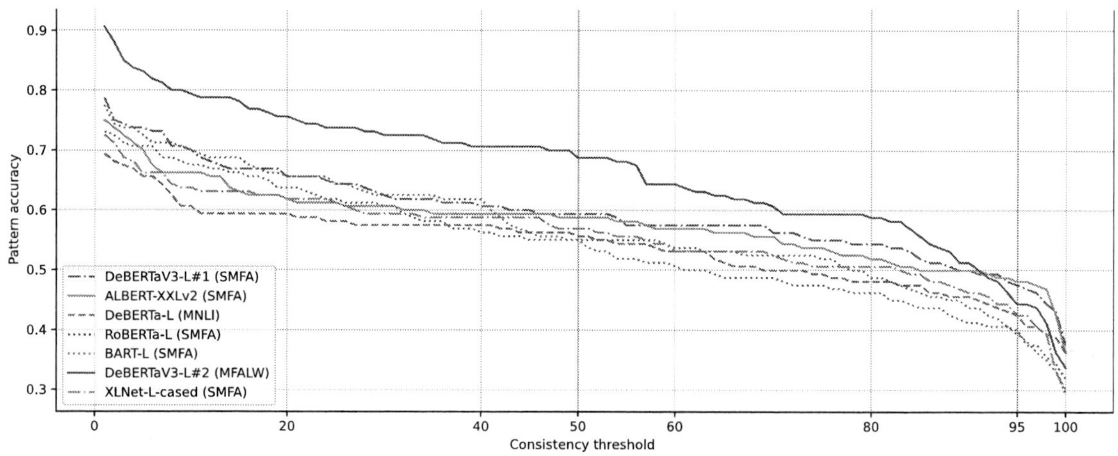

Figure 6: Pattern accuracy curves of the NLI models from Table 4. The area under the curve represents the standard accuracy based on the NLI problems.

Does ChatGPT Resemble Humans in Processing Implicatures?

Zhuang Qiu, Xufeng Duan, Zhenguang Cai
Department of Linguistics and Modern Languages
The Chinese University of Hong Kong
{zhuangqiu, zhenguangcai}@cuhk.edu.hk
xufeng.duan@link.cuhk.edu.hk

Abstract

Recent advances in large language models (LLMs) and LLM-driven chatbots, such as ChatGPT, have sparked interest in the extent to which these artificial systems possess human-like linguistic abilities. In this study, we assessed ChatGPT's pragmatic capabilities by conducting three preregistered experiments focused on its ability to compute pragmatic implicatures. The first experiment tested whether ChatGPT inhibits the computation of generalized conversational implicatures (GCIs) when explicitly required to process the text's truth-conditional meaning. The second and third experiments examined whether the communicative context affects ChatGPT's ability to compute scalar implicatures (SIs). Our results showed that ChatGPT did not demonstrate human-like flexibility in switching between pragmatic and semantic processing. Additionally, ChatGPT's judgments did not exhibit the well-established effect of communicative context on SI rates.

1 Introduction

In recent years, large language models (LLMs) have achieved unprecedented success in various linguistic tasks, such as disambiguation (Ortega-Martín, 2023), question answering (Brown et al., 2020) and translation (Jiao et al., 2023). However, there is still ongoing debate among researchers about whether these LLMs truly approximate human cognition and language use. On the pessimistic side, Chomsky et al. (2023) argued that "[LLMs] differ profoundly from how humans' reason and use language. These differences place significant limitations on what these programs can do, encoding them with ineradicable defects". In contrast, others have taken a more optimistic view. Piantadosi (2023) argued that recent LLMs should be considered as cognitive models of how people represent and use language.

To address this ongoing debate, researchers have taken an empirical approach by subjecting LLMs to various psychological experiments. Binz and Schulz (2023) subjected GPT-3 to psychological experiments originally designed to study aspects of human cognition such as decision-making, information search and causal reasoning. They found that GPT-3 exhibited human-like or even better-than-human performance in tasks like gamble decisions and multiarmed bandit tasks, with signs of model-based reinforcement learning. Kosinski (2023) tested several language models using the false-belief tasks commonly used to test theory of mind (ToM) in humans. They found that recent GPT models, including GPT-4, GPT-3.5, and GPT-3, provided ToM-like responses similar to those of school children. However, more recent research suggests that ChatGPT's deployment of ToM was not as reliable as that of humans (Brunet-Gouet, Vidal, and Roux, 2023).

Cai et al. (2023) investigated whether ChatGPT resembles humans in language comprehension and production by conducting 12 experiments on psycholinguistic effects at different linguistic levels. They found that ChatGPT exhibited human-like patterns of language use in 10 out of the 12 experiments. For instance, in speech perception, it demonstrated sound-shape (Westbury, 2005) and sound-gender association (Cassidy, Kelly & Sharoni, 1999); in lexical processing, it updated meanings of ambiguous word according to recent input (Rodd et al., 2013); in syntactic processing, it reused recently-encountered syntactic structures (Bock, 1986); in semantic processing, it inferred

Proceedings of the 4th Natural Logic Meets Machine Learning Workshop (NALOMA23), pages 25–34
June 20, 2023.

the likelihood that a sentence is implausible as a result of noise corruption (Gibson et al., 2013) and glossed over errors; at the discourse level, it drew inferences and attributed causality of events according to verb meanings; it was also sensitive to the interlocutor in meaning access and word choice. These results demonstrate that ChatGPT is profoundly similar to humans in its language use. However, it's worth noting that ChatGPT also failed to replicate human patterns in two of the experiments. In one, while humans tend to use shorter words to express less information (e.g., Mahowald et al., 2013), ChatGPT did not display this tendency. In another, ChatGPT did not make use of context to disambiguate syntactic ambiguities (Altmann and Steedman, 1988).

As we delve deeper into LLM-human similarities, it is vital to scrutinize the degree to which ChatGPT's language use aligns with that of humans and to reflect on the implications of such similarities for the evolution of artificial intelligence. Thus, it is important that LLMs are comprehensively tested in order to evaluate how human-like their language use is. So far, one aspect of language use that has not been examined is pragmatics. A hallmark of human language is the ability to convey meanings beyond the literal meaning of the words, through the use of pragmatic implicatures (Grice, 1975; 1978). Experimental pragmatics research has shown that humans can distinguish implicatures from the literal meaning of utterances, and that the computation of implicatures is influenced by the communicative context (Doran et al., 2012; Zondervan, 2010; Bonnefon, Feeney and Villejoubert, 2009). In this project, we assessed the pragmatic capabilities of LLMs by subjecting ChatGPT to three pre-registered experiments that focused on the computation of pragmatic implicatures. The first experiment aimed to determine whether ChatGPT is able to inhibit the computation of generalized conversational implicatures (GCIs) when explicitly required to process the literal meaning of the text. The second and third experiments tested whether the communicative contexts affect how ChatGPT computes scalar implicatures (SIs).

2 Experiment 1

In this experiment, we tested whether ChatGPT can distinguish "what is said" from "what is implicated" as human beings do. According to standard linguistic accounts, "what is said" refers to the truth-conditional meaning of an utterance, while "what is implicated" refers to the pragmatic implicature, which is an additional level of meaning that is enriched during the conversation (Grice, 1975; 1978). For instance, consider the sentence "Bill caused the car to stop" (Levinson, 2000, p. 39). While this sentence is semantically compatible with the scenario in which Bill slammed on the brakes, its implicature suggests that Bill stopped the car in an unconventional way, thus excluding the possibility that he stopped it with the foot pedal.

The computation of such implicature is believed to follow general principles of conversation and involve reasoning about the possible alternatives that the speaker could have used (Grice, 1975). For example, interlocutors are expected to be truthful while also making their utterances clear and understandable. If Bill stopped the car in a typical way, the speaker would have said something like "Bill slammed on the brakes." The fact that the speaker didn't use this typical expression implies that Bill didn't use the brakes to stop the car and might have stopped it in an unconventional way. This pragmatic implicature is enriched based on the literal meaning of the utterance. We are so used to interpreting utterances pragmatically that we often bypass their literal meaning, unless the implicature is explicitly canceled, as in "Bill caused the car to stop, I mean he slammed on the brakes."

A critical question in the study of pragmatic implicatures is whether non-experts can differentiate between "what is said" and "what is implicated." To address this issue, Doran, Ward, Larson, McNabb, and Baker (2012) measured the rate at which people compute a variety of generalized conversational implicatures (GCIs) in different experimental manipulations. These GCIs are implicatures that can be inferred without reference to the context (Grice, 1975). The study found that, by default, participants were able to derive the implicature of an utterance around half the time. However, the computation of GCIs decreased if participants were explicitly instructed to focus only on the literal meaning of the utterance. This suggests that non-experts without training in linguistics can still distinguish pragmatic implicature from the literal meaning. We adopted the experimental design of Doran et al. (2012) to investigate whether ChatGPT exhibits similar patterns to human participants when processing GCIs.

2.1 Design and stimuli

The design of this experiment was based on that of Doran et al. (2012). As shown in (1), ChatGPT was presented a mini dialogue, where Irene asked a question and Sam responded to the question. The mini dialogue was followed by a statement of the fact. ChatGPT was then asked to decide, given the factual statement, whether Sam's response was true or false.

1.Q-based GCI:

Irene: How much cake did Gus eat at his sister's birthday party?
Sam: He ate most of the cake.
FACT: By himself, Gus ate his sister's entire birthday cake.

In (1), the GCI in question belongs to what is called a "Q-based" implicature (Levinson, 2000), where a weaker quantifier (i.e., "most") in the scale of informativeness implicates the negation of a stronger quantifier (i.e., "all", as expressed by the word "entire" in the factual statement). That is, quantifiers "some-most-all (entire)" form a scale of increasing informativeness in that if "all of X" holds, then "most of X" holds, and "some of X" must hold, but not vice versa. Given the scale, the utterance "some of X" implicates the negation of "most of X" and "all of/ entire X"; similarly, the utterance of "most of X" implicates the negation of "all of/ entire X". Thus, based on the factual statement, Sam's response is logically true but pragmatically infelicitous. Judging Sam's response as false indicates successful GCI computation and judging it as true indicates the computation of the literal meaning but not of GCI.

Apart from Q-based GCIs, Doran et al. (2012) also investigated two other types of GCIs: "I-based" implicatures and "M-based" implicatures. The former refers to cases where the speaker says as little as necessary while the listener needs to "amplify the informational content of the speaker's utterance by finding the most specific interpretation" (Levinson, 2000). For example, the utterance "She walked into the bathroom. The window was open." has the implicature that the window is in the bathroom, while the truth-conditional meaning of the utterance allows for the possibility that the window is located elsewhere. "M-based" implicatures refer to cases where the speaker uses a marked way in the description of a common state of affairs, implicating that the unmarked form of the state of affairs does not hold. For instance, the phrase "waited and waited" implies an extended duration of waiting, despite its literal meaning being agnostic to the length of the waiting period. The three types of GCIs each have their own subcategories, as detailed in Appendix A. Each subcategory consisted of four experimental items, resulting in a total of 44 experimental items. Additionally, 16 filler items were included (taken from Doran et al., 2012), which did not require the computation of GCIs.

The experiment had two conditions: pragmatic and literal. In the pragmatic condition, ChatGPT was instructed to evaluate the truth of Sam's response based on the factual statement. After each dialogue and the factual statement, we prompted ChatGPT with "Please judge whether what Sam says is true or false based on the fact." In the literal condition, ChatGPT was instructed to interpret Sam's response literally. We prompted ChatGPT with "Please judge whether what Sam says is literally true or false based on the fact." Doran et al. (2012) found that, compared to the literal condition, the pragmatic condition led human participants to compute more GCIs (i.e., to evaluate Sam's responses more often as false). We aimed to investigate whether ChatGPT exhibits similar sensitivity to the instructions in drawing GCIs.

2.2 Procedure

We followed the data collection procedure preregistered with the Open Science Framework (`https://osf.io/cp29j`), eliciting responses from ChatGPT (Feb 13 version)[1]. In each run, we used a Python script to simulate a human interlocutor having a conversation with ChatGPT. We first presented a training example (in the pragmatic or literal condition), followed by actual experimental stimulus (see Appendix A). ChatGPT was instructed to respond by saying only "true" or "false" without other words or explanations, and we recorded the responses. In total, this study had 400 runs, with 200 runs for each condition.

[1] The original study of Doran et al. (2012) included a third condition known as the "literal Lucy" condition, which was also included in our preregistration. We specified that we would only collect data for this condition if ChatGPT could pass a sanity check test. Our testing revealed that ChatGPT consistently failed the sanity check. As per our preregistration plan, we did not collect data for this condition.

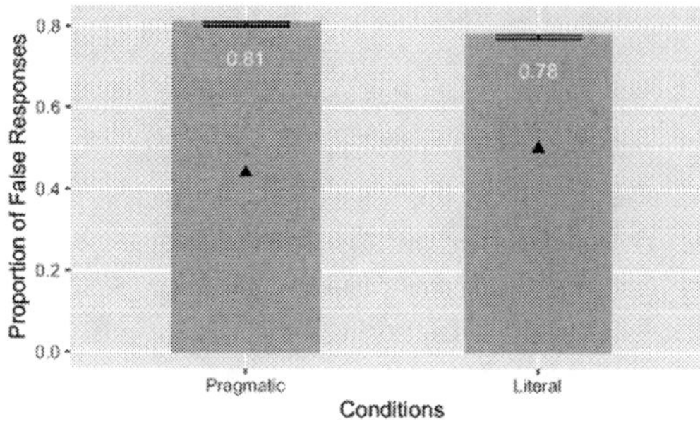

Figure 1: Proportion of false responses (i.e., GCIs) in the pragmatic and literal condition in Exp1. Note, the error bars represent confidence interval (computed using bootstrapping). The triangles represent conditional means from human participants in Doran et al. (2012).

2.3 Results and Discussion

Doran et al. (2012) found that human participants in the pragmatic condition were more likely to evaluate Sam's response as false (50%) than those in the literal condition (44%), and such a difference was statistically significant. Given that in all the experimental items, Sam's response was pragmatically infelicitous but logically compatible with the fact, the "false" judgements reflected the computation of GCIs. In this study, we found much higher rates of "false" judgements for the experimental items in both the pragmatic condition (81%) and the literal condition (78%) (see Figure 1). Following the preregistered analytical plan, we applied a Bayesian generalized linear model to trial-level responses (true or false, using true as the reference level), using condition (pragmatic vs. literal) as the predictor. The random effects structure consisted of by-item intercepts and slopes, which was the maximal random effects structure for a between-subjects design. Though there was a slight decrease of false responses in the literal compared to the pragmatic condition, this difference was not statistically significant (beta = -0.15, CI = [-0.9, 0.63]). As an exploratory analysis, we investigated the possibility that the effect of the condition was modulated by the category of the GCIs. Another Bayesian generalized linear model was constructed using the condition (pragmatic vs literal, dummy-coded with the pragmatic condition being the reference level), the category of the GCIs (I-based, M-based, and Q-based, dummy-coded with the Q-based GCIs being the reference level), and their interactions to predict the probability of giving a false response (i.e., GCI). The results showed that none of the effects in the model were statistically meaningful (see Table 1). Instead of showing human-like flexibility switching between pragmatic and semantic interpretation, ChatGPT was unable to inhibit the computation of GCIs even when it was instructed to do so.

3 Experiment 2

In this experiment, we aimed to further investigate ChatGPT's ability to draw pragmatic inferences, specifically in relation to a type of Q-based GCIs known as scalar implicatures (SIs). SIs are a well-studied phenomenon where the presence of a lower scalar item implies the negation of the higher scalar items (Horn, 1972). For instance, the sentence

	Estimate	Est.Error	l-95% CI	u-95% CI
Intercept	3.89	1.22	1.63	6.40
Literal	-0.66	0.52	-1.69	0.36
M-Based GCIs	0.07	2.04	-3.93	4.08
I-Based GCIs	0.55	1.81	-3.00	4.12
Literal:M-Based GCIs	0.94	0.96	-0.87	2.92
Literal:I-Based GCIs	0.76	0.77	-0.71	2.34

Table 2: The effect of condition, the category of the GCIs and their interactions in Exp1. Note, an estimate is statistically meaningful when zero is not included within the 95% credible interval.

"Sam had a hot dog or a hamburger for lunch" implies that Sam did not have both a hot dog and a hamburger for lunch, even though the sentence's literal meaning allows for this possibility.

Zondervan (2010) argued that an important contextual factor that influences the interpretation of scalar items is the information structure-whether the scalar item concerns the information focus or information background. For example, the sentence "Julie had found a crab or a starfish", can be the answer to two different questions as follows:

2a. What had Julie found?
2b. Who had found a crab or a starfish?

Depending on the question, the same sentence "Julie had found a crab or a starfish" has different information structure. When it is the answer to question 2a, the second half of the sentence including the scalar item "or" is the information focus (new information), while the first half of the sentence including the subject and main verb is the information background (given information). On the other hand, if the same sentence is the answer to question 2b, the subject "Julie" becomes the information focus while the scalar item retreats to the information background. Zondervan conducted a series of experiments, showing that readers are more likely to derive the SI of "or" when it is part of the information focus compared with the cases in which the scalar item is part of the information background. We wonder if ChatGPT resembles human beings showing similar sensitivity to conversational context when processing scalar item "or". If ChatGPT has acquired the pragmatic knowledge similar to that of the humans, it should be more likely to interpret the expression "A or B" as "A or B but not both A and B" when it is part of the information focus compared with the case in which the expression "A or B" is part of the information background. To further explore the way ChatGPT processes scalar items, we replicated the second experiment in Zondervan (2010) using ChatGPT as the participant.

3.1 Design and stimuli

The experimental items of the study consisted of six short story pairs, each followed by a true-or-false question. All the stories ended with a conversation between two characters, in which one character used the scalar item "or" in his/her reply to another character's question (see 3 and 4). Each story in a pair differed in terms of the context where the scalar item occurred- whether the scalar item being part of the information focus or the information background. In the scalar-implicature-relevant (SI-relevant) condition (see 3), the question was about the object ("what" question), and the scalar item "or" was part of the information focus. In this case, the interpretation of the scalar item as either "A or B but not both A and B" or "A or B and possibly both A and B" had particular relevance to the conversation. In the scalar-implicature-irrelevant (SI-irrelevant) condition (see 4), the question is about the subject ("who" question), and the scalar item was part of the information background. Thus, the interpretation of the scalar item was not the major concern of the conversation. Crucially, based on the information provided in the story, the using of the scalar item "or" was logically sound but pragmatically infelicitous, and at the end of the story, ChatGPT was asked to judge if the character's answer was true or false. If the SI of "or" was computed, ChatGPT would respond with "false" to the question; or conversely, if the SI was not computed, a "yes" judgement would be given.

3. SI-relevant:

Julie and Karin were searching for marine animals on the beach. After some searching Julie found a crab. Not much later she also found a starfish. Unfortunately, Karin didn't find anything. When Karin returned, her mother asked what kind of marine animals Julie had found. Karin answered that Julie had found a crab or a starfish.

Is Karin's answer true or false?

4. SI-irrelevant:

Julie and Karin were searching for marine animals on the beach. After some searching Julie found a crab. Not much later she also found a starfish. Unfortunately, Karin didn't find anything. When they returned, their mother asked who had found a crab or a starfish. Karin answered that Julie had found a crab or a starfish.

Is Karin's answer true or false?

In Zondervan's original study (2010), the experimental items comprised six pairs of stories similar to (3) and (4) but written in Dutch. For the present study, we utilized the English versions of these stories as the experimental items. Additionally, we created 14 filler items that mirrored the length and structure of the experimental items. Each filler item contained a dialogue in which one character answered the question posed by the other character. Half of the filler items were designed to elicit a "true" response, while the other half were designed to elicit a "false" response. To balance the experimental conditions and the order of stimuli, we employed four pseudo-randomized lists of items, following Zondervan's original study.

3.2 Procedure

We followed the data collection procedure preregistered with the Open Science Framework (`https://osf.io/egm7v`), eliciting responses from ChatGPT (Feb 13 version). In each run of the experiment, we used a Python script to simulate a human interlocutor having a conversation with ChatGPT. At the start, the human interlocutor instructed ChatGPT to make truth-value judgements based on the content of the stories. Two practice trials were given to ChatGPT, the correct answer of which was "true" and "false" respectively. After the practice trial, ChatGPT was randomly assigned to one list of items, which were presented sequentially. For each item, ChatGPT was instructed to respond by saying only "true" or "false" without other words or explanations, and we recorded the responses from ChatGPT. In total, this study had 200 runs of the script, with 50 runs for each list of items.

3.3 Results and Discussion

In Zondervan (2010), the rate of "false" judgements (i.e., SIs) was 67% in the SI-relevant condition and 41% in the SI-irrelevant condition. In our experiment, ChatGPT responded with "true" for more than 99% of the experimental items, regardless of whether the item was in the SI-relevant or SI-irrelevant condition. The "true" judgement meant that ChatGPT judged the pragmatic infelicitous usage of "or" as "true", which suggested a lack of pragmatic interpretation. Only one trial in the SI-relevant condition and two

	"False"	"True"
Experimental items		
SI-relevant	1	599
SI-irrelevant	2	598
Filler items		
Correct Answer: False	1394	6
Correct Answer: True	96	1304

Table 2: A summary of judgements from ChatGPT for experimental items and filler items across different conditions in Exp2. Note, the column labels indicate the judgements provided by ChatGPT.

trials in the SI-irrelevant condition received a "false" judgement, which was typically interpreted as the computation of SIs (see Table 2). Given the large number of trials in the experiment, the difference between SI-relevant and SI-irrelevant condition regarding the rate of SI computation was not statistically meaningful (beta = -1.31, CI = [-10.81, 4.78]).

Our analysis of the filler items revealed that ChatGPT demonstrated sensitivity to the truth conditions of the statements (see Table 2). When the character in the story provided an untruthful response, and thus the correct answer to the question should have been "false", ChatGPT provided more "false" judgments than "true" judgments (1394 vs. 6). Conversely, when the correct answer to the filler item was "true", ChatGPT provided more "true" judgments than "false" judgments (1304 vs. 96). To further explore the impact of the correct answer on ChatGPT's judgments, we modeled the probability of ChatGPT providing a "false" judgment as a function of whether the correct answer to the filler item was "true" or "false" (both dummy coded with the "false" answer being the reference level). Maximal random effects structures were constructed including subject and item intercepts and slopes. We found that when the correct answer of the filler item was "true", the "false" judgements from ChatGPT decreased at a statistically meaningful rate (beta = -19.64, CI = [-33.92, -11.66]). In total, the accuracy rate of ChatGPT in answering the filler items was above 85 percent.

In this experiment, we investigated whether ChatGPT exhibited human-like patterns of scalar implicature computation by responding to the information structure of the communicative context. Previous research on human participants has shown that when the scalar item "or" was in the information focus, they were more likely to derive the upper bounded reading ("A or B but not both A and B") compared to when the scalar item was in the information background. Our findings suggest that ChatGPT consistently provided "true" responses when asked if "A or B" is true when both A and B occur, indicating that it interpreted the scalar item "or" as lower bounded ("A or B and possibly both A and B") for over 99% of the trials, regardless of whether it appeared in the information focus or background. Furthermore, ChatGPT did not always provide "true" responses. For filler items where the correct answer was "false", ChatGPT provided significantly more "false" responses than "true" responses, and its accuracy rate was high. Therefore, the reason why ChatGPT almost always provided a "true" response for experimental items was that it always endorsed the pure logical interpretation rather than the pragmatic interpretation of the scalar item "or". The lack of scalar implicature computation for this scalar item and the insensitivity to the information structure of the communicative context differentiate ChatGPT from human participants.

4 Experiment 3

For human participants, the computation of SI is modulated by the conversational context, and the result of Experiment 2 suggested that ChatGPT lacked the sensitivity to the manipulation of information structure, an important aspect of the conversational context. This experiment aimed to investigate whether conversational context affects how ChatGPT processes scalar implicature (SI) using a different contextual aspect and a different scalar item. Bonnefon, Feeney, and Villejoubert (2009) found that the rate of endorsing SIs for the scalar item "some" decreased when the lower bounded interpretation ("some and possibly all") threatened the face of the listener, compared to when it boosted the listener's face. In this experiment, we aimed to test whether ChatGPT shows similar sensitivity to conversational context. We adopted the same design as the first study in Bonnefon, Feeney, and Villejoubert (2009), comparing the rate of SI computation across two within-participants conditions. Unlike the original study, we did not recruit human participants but tested whether ChatGPT exhibits similar performance as human participants. Specifically, we examined whether ChatGPT is more likely to interpret the scalar item "some" as "some but not all" in the face-boosting context, but not so much when the scalar item "some" appears in the face-threatening context.

4.1 Design and stimuli

In this experiment, ChatGPT read two scenarios which were either face-threatening or face-boosting, and the scalar item "some" appeared in the description of the scenario. After reading each scenario, ChatGPT was required to answer a yes-no question. Specifically, we asked ChatGPT whether it would endorse the lower-bounded interpretation of some (which is "some and possibly all"). An example of the experimental item in the face-threatening and face-boosting context was shown in (5) and (6):

5. Face-threatening context:

Imagine that you have joined a poetry club, which consists of five members in addition to you. Each week, one member writes a poem, and the five other members discuss the poem in the absence of its author. This week, it is your turn to write a poem and to let others discuss it. After the discussion, one fellow member confides to you that "Some people hated your poem."

Yes/No question: From what this fellow member told you, do you think it is possible that everyone hated your poem?

6. Face-boosting context:

Imagine that you have joined a poetry club, which consists of five members in addition to you. Each week, one member writes a poem, and the five other members discuss the poem in the absence of its author. This week, it is your turn to write a poem and to let others discuss it. After the discussion, one fellow member confides to you that "Some people loved your poem."

Yes/No question: From what this fellow member told you, do you think it is possible that everyone loved your poem?

We included two scenarios like 5 and 6, creating two lists of items using the Latin Squared Design. All items in the experiment were directly adopted from Bonnefon, Feeney and Villejouber (2009).

4.2 Procedure

We followed the data collection procedure preregistered with the Open Science Framework (`https://osf.io/3v9gn`), eliciting responses from ChatGPT (Feb 13 version). In each run of the experiment, we used a Python script to simulate a human interlocutor having a conversation with ChatGPT. At the start, the human interlocutor instructed ChatGPT to answer yes-no questions based on the description of scenarios. Two practice trials were given to ChatGPT, the correct answer of which was "yes" and "no" respectively. After that, ChatGPT was randomly assigned to one list of items, which were presented to ChatGPT in a random order. For each item, ChatGPT was instructed to respond by saying only "yes" or "no" without other words or explanations, and we recorded the responses from ChatGPT. In total, this study had 200 runs of the script, with 100 runs for each list of items.

	"No"	"Yes"
Face-boosting	198	0
Face-threatening	198	0

Table 3: A summary of judgements from ChatGPT for experimental items across different conditions in Exp3.

4.3 Results and Discussion

According to our preregistered data exclusion criteria, we excluded data from two runs of the experiment because ChatGPT answered the second practice trial incorrectly, indicating that it may not provide reliable judgments in that run of the experiment. Therefore, we analyzed the data from 198 runs of the experiment. In Bonnefon, Feeney and Villejouber's (2009) study, 83% of human participants responded with "no" when asked if the lower bounded interpretation of "some" was possible in the face-boosting context, while a significantly lower 58% responded "no" in the face-threatening context. In contrast, our study found that ChatGPT always responded "no" to all of the trials, regardless of whether the context was face-boosting or face-threatening (see Table 3).

Though the exact mechanism is still unclear regarding why human participants were more likely to interpret the construction "some verb-ed X" as "some and possibly all verb-ed X" in the face threatening context than in the face boosting context, Bonnefon, Feeney and Villejouber (2009) suggested that the listener may take into account the intension of the speaker to use the word "some" in an underinformative way in order to protect the face of the listener. Although, the SI rate of "some" decreased in the face threatening condition, in general, human participants preferred the pragmatic interpretation of "some" as "some but not all", and that is why even in the face-threatening condition, the majority of the human participants (58%) provided a "no" judgement to the question "Do you think it is possible that everyone hated…" In our experiment with ChatGPT, we clearly saw a stronger preference for the pragmatic interpretation of "some" over the truth-conditional interpretation. In fact, ChatGPT exhibited zero variance in its judgements- for all the trials that contained the scalar item "some", ChatGPT always interpreted them as "some but not all", and thus said "no" to the question, regardless of whether the implicature was face threatening or face boosting to the listener.

5 General Discussion and Conclusion

In three experiments, we investigated whether LLMs like ChatGPT exhibit human-like performance when processing pragmatic implicatures. Previous research has shown that humans distinguish implicatures from the truth-conditional meaning of the utterance, and several factors have been identified that modulate the probability of implicature computation. While pragmatic enrichment is an essential component of successful communication, whether an implicature is computed by a specific listener in a specific communicative context is probabilistic in nature. In contrast, our findings revealed that ChatGPT lacked human-like flexibility in switching between pragmatic and semantic interpretation, as it was unable to inhibit the computation of GCIs even when instructed to do so. Notably, the processing of scalar items in ChatGPT exhibited a deterministic pattern: whereas "some" always received an upper bounded interpretation as "some but not all", the expression "A or B" almost always received a lower bounded interpretation as "A or B and possibly both A and B".

Given ChatGPT's impressive human-like performance across a range of language tasks (Cai et al., 2023), one might question why humans and LLMs differ in their computation of GCIs. Our argument is that this difference can be explained by the acquisition of GCIs and the computational resources available to humans and machines. Developmental research indicates that scalar items are acquired with a lower bounded interpretation before pragmatic enrichments (Noveck, 2001). Consequently, adults have access to both the literal and pragmatic interpretations of a scalar item, whereas LLMs are exposed to language data that are mainly pragmatically driven. This explains why ChatGPT, in general, is more prone to pragmatic interpretation compared with human participants. However, it is still unclear why some specific word like "or" almost always evokes a literal rather than pragmatic interpretation. Furthermore, humans possess limited computational resources compared to machines. The principle of economy suggests that the human mind enriches the truth-conditional meaning only when the context necessitates it (Noveck & Sperber, 2007). This echoes the fact that the effect of contextual manipulation has only been observed among human participants rather than LLMs. It is consistent with the observation that humans tend to use shorter forms of words (e.g., math instead of mathematics) when the meaning is predictable, while ChatGPT does not (Cai et al., 2023). Overall, our experiments demonstrate that although LLM-based chatbots such as ChatGPT excel in many language tasks, they do not mimic humans in their computation of GCIs.

Limitations

The scope of our research is limited to uncovering the distinction between humans and LLMs in a specific aspect of pragmatic processing: the computation of GCIs. While we offer tentative explanations for the patterns we observed, our study does not directly provide solutions for improving the performance of LLMs. In this study, we use ChatGPT as an example of LLMs due to its prominence in current research. However, it remains uncertain whether other LLMs exhibit comparable characteristics and tendencies as observed in ChatGPT. Moreover, it is important to note that our findings may not generalize to the processing of other types of pragmatic implicatures.

References

Gerry Altmann, and Mark Steedman.1988. Interaction with context during human sentence processing. *Cognition* 30, no. 3: 191-238.

Marcel Binz and Eric Schulz. 2023. Using cognitive psychology to understand GPT-3. *Proceedings of the National Academy of Sciences* 120, no. 6: e2218523120.

J. Kathryn Bock. 1986. Syntactic persistence in language production. *Cognitive psychology* 18, no. 3: 355-387.

Jean-François Bonnefon, Aidan Feeney, and Gaëlle Villejoubert. 2009. When some is actually all: Scalar inferences in face-threatening contexts. *Cognition* 112, no. 2: 249-258.

Tom Brown, Benjamin Mann, Nick Ryder, Melanie Subbiah, Jared D. Kaplan, Prafulla Dhariwal, Arvind Neelakantan et al. 2020. Language models are few-shot learners. *Advances in neural information processing systems* 33: 1877-1901.

Eric Brunet-Gouet, Nathan Vidal, and Paul Roux. 2023. Do conversational agents have a theory of mind? A single case study of ChatGPT with the Hinting, False Beliefs and False Photographs, and Strange Stories paradigms. *HAL Open Science.* https://hal.science/hal-03991530/

Zhenguang G Cai, David A. Haslett, Xufeng Duan, Shuqi Wang, and Martin J. Pickering. 2023. Does ChatGPT resemble humans in language use? *arXiv preprint* arXiv:2303.08014.

Kimberly Wright Cassidy, Michael H. Kelly, and Lee'at J. Sharoni. 1999. Inferring gender from name phonology. *Journal of Experimental Psychology: General* 128, no. 3 (1999): 362.

Noam Chomsky, Ian Roberts, and Jeffrey Watumull. 2023. The false promise of ChatGPT. *The New York Times*.https://www.nytimes.com/2023/03/08/opinion/noam-chomsky-chatgpt-ai.html

Ryan Doran, Gregory Ward, Meredith Larson, Yaron McNabb, and Rachel E. Baker. 2012. A novel experimental paradigm for distinguishing between what is said and what is implicated. *Language*: 124-154.

Edward Gibson, Leon Bergen, and Steven T. Piantadosi. 2013. Rational integration of noisy evidence and prior semantic expectations in sentence interpretation. *Proceedings of the National Academy of Sciences* 110, no. 20: 8051-8056.

Herbert Paul Grice. 1975. Logic and conversation. *In Speech acts*, pp. 41-58. Brill.

Herbert Paul Grice. 1978. Further notes on logic and conversation. In *Pragmatics*, pp. 113-127. Brill

Laurence Robert Horn. 1972. *On the semantic properties of logical operators in English.* University of California, Los Angeles.

Wenxiang Jiao, Wenxuan Wang, Jen-tse Huang, Xing Wang, and Zhaopeng Tu. 2023. Is ChatGPT a good translator? A preliminary study. *arXiv preprint* arXiv:2301.08745.

Michal Kosinski. 2023. Theory of mind may have spontaneously emerged in large language models. *arXiv preprint* arXiv:2302.02083.

Stephen C Levinson. 2000. *Presumptive meanings: The theory of generalized conversational implicature.* MIT press.

Kyle Mahowald, Evelina Fedorenko, Steven T. Piantadosi, and Edward Gibson. 2013. Info/information theory: Speakers choose shorter words in predictive contexts. *Cognition* 126, no. 2 (2013): 313-318.

Ira A Noveck. 2001. When children are more logical than adults: Experimental investigations of scalar implicature. *Cognition* 78, no. 2: 165-188.

Ira A Noveck and Dan Sperber. 2007. The why and how of experimental pragmatics: the case of 'scalar inferences, in *Advances in Pragmatics*, ed N. Burton-Roberts (Basingstoke: Palgrave), 184–212.

Miguel Ortega-Martín, Óscar García-Sierra, Alfonso Ardoiz, Jorge Álvarez, Juan Carlos Armenteros, and Adrián Alonso. 2023. Linguistic ambiguity analysis in ChatGPT. *arXiv preprint* arXiv:2302.06426

Steven T Piantadosia. 2023. Modern language models refute Chomsky's approach to language. *Lingbuzz Preprint*, lingbuzz/007180

Jennifer M. Rodd, Belen Lopez Cutrin, Hannah Kirsch, Alessandra Millar, and Matthew H. Davis. 2013. Long-term priming of the meanings of ambiguous words. *Journal of Memory and Language* 68, no. 2 (2013): 180-198.

Chris Westbury. 2005. Implicit sound symbolism in lexical access: Evidence from an interference task. *Brain and language* 93, no. 1 (2005): 10-19.

Arjen Zondervan. 2010. *Scalar implicatures or focus: an experimental approach.* Netherlands Graduate School of Linguistics.

A Appendix

An example of experimental items containing GCIs of different categories in Exp1.

Dialogue	Fact	First Level Category	Second Level Category
Irene: Hey, Sam. Do you know who wrote Pride and Prejudice? Sam: A British woman wrote it, and her last name was Austen.	FACT: Jane Austen, a British woman, wrote Pride and Prejudice.	Training Example	Training Example
Irene: How much cake did Gus eat at his sister's birthday party? Sam: He ate most of the cake.	FACT: By himself, Gus ate his sister's entire birthday cake.	Q_Based_GCIs	Quantifiers_Modals
Irene: How many children does Lisa have? Sam: Lisa has three children.	FACT: Lisa has quadruplets	Q_Based_GCIs	Cardinals
Irene: How would you say you're doing financially? Sam: I'm comfortable.	FACT: Sam just bought four condos at Lake Point Tower, in downtown Chicago, where Oprah Winfrey lives.	Q_Based_GCIs	Gradable_Adjectives
Irene: What kind of milk does your diet allow for? Sam: It allows for 1%.	FACT: The only type of milk prohibited by Sam's diet is full-fat milk.	Q_Based_GCIs	Rankings
Irene: I heard something big happened in the art studio yesterday. Sam: In a fit of rage, Rachel picked up a hammer and broke a statue.	FACT: After grabbing a hammer, Rachel angrily kicked a statue, causing it to fall over and break.	I_Based_GCIs	Argument_Saturation
Irene: What happened when Sue came over? Sam: She walked into the bathroom. The window was open.	FACT: The open windows are in the kitchen, and there are no windows in the bathroom.	I_Based_GCIs	Bridging_Inferences
Irene: Can the guys come to the reception? Sam: George and Steve play squash at the gym until 6:00 every day.	FACT: George plays squash at the YMCA until 6:00 daily, and Steve plays squash at SPAC until 6:00 every day.	I_Based_GCIs	Coactivities
Irene: I understand that George has had a really rough year. Sam: Last month, he lost his job and started drinking.	FACT: George started drinking on the 15th of last month and lost his job on the 20th of last month.	I_Based_GCIs	Conjunction_Buttressing
Irene: Why is Stephen so upset? Sam: He caused Bill to die.	FACT: Stephen intentionally murdered Bill.	M_Based_GCIs	Verbal_Periphrasis
Irene: What happened at Doctor Witherspoon's office? Sam: Sasha waited and waited for her appointment.	FACT: Sasha waited 5 minutes for her appointment at DoctorWitherspoon's office.	M_Based_GCIs	Repeated_Verb_Conjuncts
Irene: What did Joseph do after finishing the marathon? Sam: He drank bottles and bottles of water.	FACT: Joseph drank one 20 oz bottle and one 16 oz bottle of water after finishing themarathon.	M_Based_GCIs	Repeated_Noun_Conjuncts

Recurrent Neural Network CCG Parser

Sora Tagami
Ochanomizu University
tagami.sora@is.ocha.ac.jp

Daisuke Bekki
Ochanomizu University
bekki@is.ocha.ac.jp

Abstract

The two contrasting approaches are end-to-end neural NLI systems and linguistically-oriented NLI pipelines consisting of modules such as neural CCG parsers and theorem provers. The latter, however, faces the challenge of integrating the neural models used in the syntactic and semantic components. RNNGs are frameworks that can potentially fill this gap, but conventional RNNGs adopt CFG as the syntactic theory. To address this issue, we implemented RNN-CCG, a syntactic parser that replaces CFG with CCG. We then conducted experiments comparing RNN-CCG to RNNGs with-/without POS tags and evaluated their behavior as a first step towards building an NLI system based on RNN-CCG.

1 Introduction

Over the years, two contrasting approaches to natural language inference (NLI) have emerged: end-to-end neural NLI systems based on large language models (LLMs) (Lan et al., 2020; Raffel et al., 2020; He et al., 2021), which we call *mono-modular* approaches, and linguistically-oriented NLI pipelines consisting of syntactic parsers, semantic representations and theorem provers (Bos et al., 2004; Chatzikyriakidis and Luo, 2014; Mineshima et al., 2015; Abzianidze, 2015; Martínez-Gómez et al., 2017; Chatzikyriakidis and Bernardy, 2019), which we call *multi-modular approaches*. While the former has become more popular in recent years and has shown remarkable progress with the increasing scale of LLMs, the latter offers high precision, explanatory properties and strength in higher-order reasoning such as arithmetic. Both approaches have strengths and weaknesses and are expected to complement each other.

A drawback of using neural networks in multi-modular approaches is that their neural models are split between syntax and semantics. For example, the neural part-of-speech (POS) taggers cannot receive feedback from the results of the semantic component. The distributional representations in semantic components considered in works such as Cooper (2019); Larsson (2020); Bekki et al. (2022, 2023) are not connected to syntax. This gap between syntactic and semantic neural models is a potential weakness of multi-modular approaches compared to mono-modular approaches that seek to optimize the whole process of NLI.

The use of Recurrent Neural Network Grammars (RNNGs) (Dyer et al., 2016) is a potential solution to bridge the gap between syntactic and semantic neural models in multi-modular approaches. RNNGs provide syntactic parsers that can function as feeding input (syntactic structures) to semantic components, which is still a non-trivial task for large language models. Furthermore, unlike standard syntactic parsers and large language models, RNNGs provide embedded representations for *phrasal* constituents obtained by training on predicting syntactic structures, which we expect to be useful in a semantic component as well.

One remaining challenge is that the underlying grammar of the current RNNGs is context-free grammar (CFG), while modern syntactic processing in the multi-modular approaches adopts mildly context-sensitive grammars such as combinatory categorial grammar (CCG) (Steedman, 1996, 2000).

Therefore, in this study, we attempt to implement *RNN-CCG*, a syntactic parser that replaces the underlying grammar of RNNGs from CFG to CCG, and compare the performance of RNN-CCG with RNNGs, as a first step towards developing a complete NLI system using RNN-CCG. Technically, RNN-CCG can be built using almost the same techniques as RNNGs, but we will show that its performance is slightly better than RNNGs.

2 Recurrent Neural Network Grammars

RNNGs are language models and syntactic parsers that explicitly model hierarchical structures of

Proceedings of the 4th Natural Logic Meets Machine Learning Workshop (NALOMA23), pages 35–40
June 20, 2023.

words and phrases. Here, we will give an example of their behavior as syntactic parsers. Internally, RNNGs use two data structures: Stack and Buffer. Initially, Buffer contains all the word vectors. Operations on them are defined as Actions:

SHIFT Pop the word vector from Buffer and push it to Stack.

NT X Push a vector corresponding to the non-terminal symbol X to Stack. This non-terminal symbol X is marked as *open*.

REDUCE Pop from Stack all the elements up to the first *open* non-terminal symbol X encountered. Generate a new vector that encodes them and push it back to Stack as a new element.

At each time step, Stack, Buffer, and history of Actions are encoded using LSTMs and RNNs. Parsing is performed by determining the next Action at each parsing state based on this encoding. It is inefficient to recompute the encoding of Stack every time; thus, RNNGs adopts a mechanism called Stack LSTMs (Dyer et al., 2015) for optimization.

RNNGs have been the subject of subsequent researches: stack-only RNNGs (Kuncoro et al., 2017), which eliminate Buffer from the architecture and use only Stack, a Pytorch implementation model (Noji and Oseki, 2021) that enables parallel execution and learning of larger data, and a model that uses Transformer instead of RNNs (Sartran et al., 2022; Qian et al., 2021). However, in this paper, we focus on comparing CFG and CCG as underlying syntactic theories of RNNGs, adopting the simplest model presented in the original paper (Dyer et al., 2016) and conducting experiments focusing on the parsing aspect.

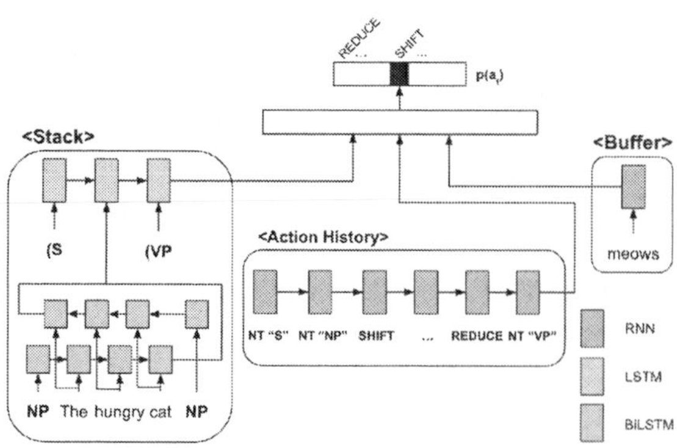

Figure 1: Architecture of RNNGs

3 RNN-CCG

We implemented two models based on RNNGs: RNN-CFG, which is a re-implementation of RNNGs using CFG, and RNN-CCG, which uses CCG instead of a CFG for the grammar used in RNNGs. By treating CCG syntactic categories as CFG terminal symbols, CCG is regarded as an instance of phrase-structure grammar, and the action selection is a multi-class classification task similar to the case of RNN-CFG. However, there was a problem with RNN-CCG in that its syntactic structures do not provide a layer for POS tags, which is insufficient to be used for semantic composition. Therefore, in this research, we extend RNN-CCG so that the structures have syntactic categories corresponding to words. For the sake of comparison, we also implement RNN-CFG that outputs non-terminal symbols corresponding to words.

3.1 Combinatory Categorial Grammar

CCG is a lexicalized grammar, the generative capacity of which is known to be mildly context-sensitive. In phrase structure grammars such as CFGs, most of the syntactic information is described by production rules, and the lexicon is relatively simple. In lexicalized grammars, on the other hand, most of the syntactic information is stored in the lexicon, and the combinatory rules are relatively simple. Additionally, CCG provides semantic information so that the syntactic structures determine the paths for semantic composition.

To generate *The hungry cat meows* in a context-free grammar, the following rules are required[1]:

$$\begin{array}{rcl} S & \rightarrow & NP\ VP \\ VP & \rightarrow & meows \\ NP & \rightarrow & The\ hungry\ cat \end{array}$$

In contrast, in CCG, each lexical item is defined as follows. In this example, two lexical items are combined using the backward function application rule, a combinatory rule in CCG, to generate a sentence.

$$\begin{array}{rcl} The\ hungry\ cat & \vdash & NP \\ meows & \vdash & S\backslash NP \end{array}$$

3.2 Part-of-speech Tags

RNNGs' syntactic structures do not contain POS tags; therefore, when implementing RNN-CCG

[1]The original paper on RNNGs does not consider the POS tags of each word; we follow this convention in this paper.

within the same framework as Dyer et al. (2016), a syntactic structure such as the one shown in Figure 2 is obtained from the output Action sequence. In Figure 2, it can be inferred that the syntactic category of *disclosed* is $S[pss]\backslash NP$, but to obtain the advantage of CCG syntactic structures, which is the path for semantic composition, it is necessary to supplement such syntactic categories of words and restore the detailed syntactic information. For example, it is unclear how to supplement the syntactic category of *were* or *'nt* in Figure 2. Therefore, in this study, as shown in Figure 3, we also make our RNN-CCG predict the category corresponding to each word using the "NT X" action.

```
(S[dcl]
 (S[dcl]
  (NP Terms)
   (S[dcl]\NP
    ((S[dcl]\NP) / (S[pss]\NP) were 'nt)
    disclosed ) )
 . )
```

Figure 2: Part-of-speech-insensitive parse tree

```
(S[dcl]
 (S[dcl]
  (NP
   (N Terms ) )
  (S[dcl]\NP
   ((S[dcl]\NP)/(S[pss]\NP)
    ((S[dcl]\NP)/(S[pss]\NP) were )
    ((S\NP)\(S\NP) n't ) )
   (S[pss]\NP disclosed ) ) )
 (. . ) )
```

Figure 3: Part-of-speech-sensitive parse trees

Commonly, several constraints are imposed in RNNGs (Dyer et al., 2016) to ensure the generation of well-formed constituent structures. In this study, we added POS tags and implemented the following constraints accordingly.

- SHIFT is immediately after "NT X"
- Always REDUCE immediately after SHIFT

These two rules mean that every single terminal symbol is reduced to a non-terminal. This non-terminal corresponds to the POS tag associated with the terminal.

4 Experiment

4.1 Experimental Setup

We implemented RNN-CFG and RNN-CCG for English using hasktorch[2], the Haskell interface for Torch. For training, we used Penn Treebank[3] as CFG data and CCGbank[4] as CCG data. We used sections 2-21 for training, section 24 for validation, and section 23 for evaluation in both corpora. Details are shown in Table 1.

Table 1: Corpus Statistics

	PTB		CCGbank	
	train	test	train	test
Sentences	39,832	2,416	39,604	2,407
Tokens	44,987	8,461	44,211	8,393
Actions(Without POS)	1,182	236	810	258
Actions(With POS)	1,229	282	1,642	542

4.2 Experimental Results

We show the micro F1 score for each model when this model is considered a sequence labeling model that predicts the actions in Table 2.[5]

Table 2: Experimental Results

	RNN-CFG		RNN-CCG	
	Without POS	With POS	Without POS	With POS
micro F1	90.7	91.3	91.3	93.6

Following the previous studies, these F1 scores are calculated for the predictions assuming that all the predictions before that time step coincide with the ground truth data.

4.3 Discussion

POS tags According to Table 2, the POS-tagged models achieved higher scores in both the CFG and CCG models. This is a welcome result given the usefulness of POS tags in semantic composition. On the other hand, this seems counter-intuitive since Table 1 shows that POS-tagged models have more actions than their untagged counterparts in both RNN-CFG and RNN-CCG. Predicting POS tags becomes more difficult when the number of classes increases in multi-class classification tasks.

This seemingly contradictory result can be attributed to the constraints discussed in Section 3.2. While the POS-free models predict a SHIFT action to move a word from Buffer to Stack, the POS-tagged models have to predict three actions in a row: "NT X", SHIFT, and REDUCE. Due to the constraints mentioned earlier, all three predictions

[2] `http://hasktorch.org/`

[3] `https://catalog.ldc.upenn.edu/LDC99T42`

[4] `https://catalog.ldc.upenn.edu/LDC2005T13`

[5] We used rt_G.large (NVIDIA V100 for NVLink 16GiB HBM2) on the ABCI (AI Bridging Cloud Infrastructure) (`https://abci.ai/`) as the experimental environment.

are guaranteed to be correct, contributing to the F1 score.

RNN-CCG vs. RNN-CFG Both the POS-tagged and POS-free RNN-CCG models outperformed their RNN-CFG counterparts in terms of microF1 score. Considering only the results of the POS-free models, attributing the differences in accuracy to the number of actions, the POS-tagged models in our study had more actions in the CCG case. Therefore, other factors must be at play. One possible explanation is that there are fewer combinatory rules in CCG grammar compared to CFG grammar. This results in a smaller pool of categories to predict with the "NT X" action, which may improve performance.

4.4 Error Analysis

In the above results, all predictions up to each time step used the ground-truth data, but when used as a parser, the prediction at each time step depends on the previous predictions. Therefore, we conducted an error analysis using the predicted results of the evaluation data by the syntax parser, including the state of Stack.

In RNN-CCG with POS, it often occurred that the same category was output repeatedly, as shown in Figure 4.

```
(S[dcl]\NP
   (S[dcl]\NP
       (S[dcl]\NP
           (S[dcl]\NP
               (S[dcl]\NP
                  ...
                  (S[dcl]\NP general )
```

Figure 4: Output of RNN-CCG with POS

This was not observed in RNN-CFG or RNN-CCG without POS. One possible cause is that many training data repeat predicting the same syntactic category during training. This is not the case in CFG, where there are not many production rules that predict the same nonterminal successively in the form of $X \to X, \ldots$. In CCG, however, this occurs when X and Y are the same in the backward function application rule. A typical example is $S\backslash NP \Rightarrow S\backslash NP, (S\backslash NP)\backslash(S\backslash NP)$, which occurs in a structure where an intransitive verb is followed by a VP modifier. While "NT $S[dcl]\backslash NP$" is continuously predicted, an intransitive verb continuously stays at the beginning of Buffer. So to learn when to transition to the SHIFT action, information about whether an adverbial phrase exists in the Buffer must be referred to.

There are also benefits to using CCG. In RNN-CFG, since the production rules are not defined in advance, there is no sense to ask which CFG rule is *correct*. Figures 5, 6, 7, and 8 are the predicted results for the same sentence by RNN-CFG and RNN-CCG, both of which are predicted incorrectly, but in Figure 6, there is no rule in CCG that has S[dcl] as the child and S[dcl] as the parent, so it is possible to judge whether the output tree is consistent according to CCG theory.

```
(NP
 (N
  (N/N
   ((N/N)/(N/N) 10)
   (N/N 1\/2 ) )
  (N
   (N % )
   (. . ) ) ) )
```

Figure 5: Correct

```
(S[dcl]
 (S[dcl]
  (NP
   (N
    (N/N 10 )
   (N
     (N/N 1\/2 )
     (N % ) ) ) )
  (. . ) ) )
```

Figure 6: Prediction by RNN-CCG

```
(NP
  (QP
    (CD 10 )
    (CD 1\/2 ) )
  (NN % )
  (. . ) )
```

Figure 7: Correct

```
(S
  (NP-SBJ
    (NP
      (NNP 10 )
      (NNP 1\/2 )
      (NNP % ) ) )
  (. . ) )
```

Figure 8: Prediction by RNN-CFG

5 Conclusion

In this study, we implemented RNN-CCG, a syntactic parser in which the grammar used inside the RNNGs was replaced from CFG to CCG, and conducted comparative experiments with RNN-CFG, a reimplementation of classical RNNGs. We also implemented their extensions with POS tags considering syntactic categories corresponding to words.

The results showed that the implementation of RNN-CCG achieved a higher F1 score than RNN-CFG with respect to the prediction of actions. Moreover, both models function effectively when considering POS tags, providing a better interface for semantic composition in the case of RNN-CCGs.

Overall, RNN-CCG is a prospective candidate of syntactic parsers in a modular NLI approach that bridges the gap between neural networks within CCG parsers and semantic modules. Future research could investigate the fusion of RNN-CCG with semantic composition and logical reasonings.

Acknowledgments

We thank the two anonymous reviewers of NALOMA'23 for their insightful comments. This work is partially supported by JST CREST Grant Number JPMJCR20D2, Japan, and JSPS KAKENHI Grant Number JP23H03452, Japan.

References

Lasha Abzianidze. 2015. Towards a wide-coverage tableau method for natural logic. In T. Murata, Koji Mineshima, and Daisuke Bekki, editors, *New Frontiers in Artificial Intelligence: JSAIisAI 2014 Workshops, LENLS, JURISIN, and GABA, Revised Selected Papers. Lecture Notes in Computer Science, volume 9067*, pages 66–82.

Daisuke Bekki, Ribeka Tanaka, and Yuta Takahashi. 2022. Learning knowledge with neural dts. In *the 3rd Natural Logic Meets Machine Learning (NALOMA III), pp.17-25, Galway, Ireland, Association of Computational Linguistics*, pages 17–25.

Daisuke Bekki, Ribeka Tanaka, and Yuta Takahashi. 2023. Integrating deep neural network with dependent type semantics. In Roussanka Loukanova, Peter LeFanu Lumsdaine, and Reinhard Muskens, editors, *Logic and Algorithms in Computational Linguistics 2021 (LACompLing2021)*, Studies in Computational Intelligence 1081, page 261–284. Springer.

Johan Bos, Stephen Clark, Mark J. Steedman, James R. Curran, and Julia Hockenmaier. 2004. Wide-coverage semantic representations from a ccg parser. In *COLING '04*.

Stergios Chatzikyriakidis and Jean-Philippe Bernardy. 2019. A wide-coverage symbolic natural language inference system. In *Proceedings of the 22nd Nordic Conference on Computational Linguistics, NoDaLiDa 2019, Turku, Finland, September 30 - October 2, 2019*, pages 298–303. Linköping University Electronic Press.

Stergios Chatzikyriakidis and Zhaohui Luo. 2014. Natural language inference in coq. *J. Log. Lang. Inf.*, 23(4):441–480.

Robin Cooper. 2019. Representing types as neural events. *Journal of Logic, Language and Information*, 28:131–155.

Chris Dyer, Miguel Ballesteros, Wang Ling, Austin Matthews, and Noah A. Smith. 2015. Transition-based dependency parsing with stack long short-term memory. *CoRR*, abs/1505.08075.

Chris Dyer, Adhiguna Kuncoro, Miguel Ballesteros, and Noah A. Smith. 2016. Recurrent neural network grammars. In *Proceedings of the 2016 Conference of the North American Chapter of the Association for Computational Linguistics: Human Language Technologies*, pages 199–209.

Pengcheng He, Xiaodong Liu, Jianfeng Gao, and Weizhu Chen. 2021. DEBERTA: Decoding-Enhanced BERT with disentangled attention. In *International Conference of Learning Representations (ICLR2021)*.

Adhiguna Kuncoro, Miguel Ballesteros, Lingpeng Kong, Chris Dyer, Graham Neubig, and Noah A. Smith. 2017. What do recurrent neural network grammars learn about syntax? In *Proceedings of the 15th Conference of the European Chapter of the Association for Computational Linguistics: Volume 1, Long Papers*, pages 1249–1258.

Zhenzhong Lan, Mingda Chen, Sebastian Goodman, Kevin Gimpel, Piyush Sharma, and Radu Soricut. 2020. ALBERT: A Lite BERT for self-supervised learning of language representations. In *International Conference of Learning Representations (ICLR2020)*.

Staffan Larsson. 2020. Discrete and probabilistic classifier-based semantics. In *the Probability and Meaning Conference (PaM 2020)*, pages 62–68. Association for Computational Linguistics.

Pascual Martínez-Gómez, Koji Mineshima, Yusuke Miyao, and Daisuke Bekki. 2017. On-demand injection of lexical knowledge for recognising textual entailment. In *Proceedings of the 15th Conference of the European Chapter of the Association for Computational Linguistics: Volume 1, Long Papers*, pages 710–720.

Koji Mineshima, Pascual Martínez-Gómez, Yusuke Miyao, and Daisuke Bekki. 2015. Higher-order logical inference with compositional semantics. In *Conference on Empirical Methods in Natural Language Processing (EMNLP2015)*, pages 2055–2061.

Hiroshi Noji and Yohei Oseki. 2021. Effective batching for recurrent neural network grammars. *CoRR*, abs/2105.14822.

Peng Qian, Tahira Naseem, Roger Levy, and Ramón Fernandez Astudillo. 2021. Structural guidance for transformer language models. In *Proceedings of the 59th Annual Meeting of the Association for Computational Linguistics and the 11th International Joint Conference on Natural Language Processing (Volume 1: Long Papers)*, pages 3735–3745.

Colin Raffel, Noam Shazeer, Adam Roberts, Katherine Lee, Sharan Narang, Michael Matena, Yanqi Zhou, Wei Li, and Peter J. Liu. 2020. Exploring the limits of transfer learning with a unified text-to-text transformer. *Journal of Machine Learning Research*, 21:1–67.

Laurent Sartran, Samuel Barrett, Adhiguna Kuncoro, Miloš Stanojević, Phil Blunsom, and Chris Dyer. 2022. Transformer grammars: Augmenting transformer language models with syntactic inductive biases at scale. *Transactions of the Association for Computational Linguistics*, 10:1423–1439.

Mark J. Steedman. 1996. *Surface Structure and Interpretation.* The MIT Press, Cambridge.

Mark J. Steedman. 2000. *The Syntactic Process (Language, Speech, and Communication).* The MIT Press, Cambridge.

TTR at the SPA: Relating type-theoretical semantics to neural semantic pointers

Staffan Larsson and **Robin Cooper**
Dept. of Philosophy, Linguistics and Theory of Science
University of Gothenburg
sl@ling.gu.se, cooper@ling.gu.se

Jonathan Ginzburg
Université Paris Cité
yonatan.ginzburg@u-paris.fr

Andy Lücking
Université Paris Cité and Goethe University Frankfurt
luecking@em.uni-frankfurt.de

Abstract

This paper considers how the kind of formal semantic objects used in TTR (a theory of types with records, Cooper, 2023) might be related to the vector representations used in Eliasmith (2013). An advantage of doing this is that it would immediately give us a neural representation for TTR objects as Eliasmith relates vectors to neural activity in his semantic pointer architecture (SPA). This would be an alternative using convolution to the suggestions made by Cooper (2019a) based on the phasing of neural activity. The project seems potentially hopeful since all complex TTR objects are constructed from *labelled sets* (essentially sets of ordered pairs consisting of labels and values) which might be seen as corresponding to the representation of structured objects which Eliasmith achieves using superposition and *circular convolution*.

1 Introduction

Work on TTR, a theory of types with records, for example Cooper (2023), claims that it can be used to model types learned by agents in order to classify objects and events in the world. If this is true, types must be represented in some way in brains. In this paper we will explore the possibility of using Eliasmith's Semantic Pointer Architecture (SPA) (Eliasmith, 2013) for this purpose. The question of neural representations of types arises in connection with the theory of types proposed by TTR in a way that it does not in connection with more traditional type theories. The reason is that TTR aims to provide the kind of types that agents use in the perception of objects and events and which they use in interaction to communicate with each other. If it were to turn out that the kind of types used are in principle impossible to represent on arrays of neurons then this would call this project into question.

We chose SPA, since it is a model of a biological neural network. Notwithstanding their practical and methodological success, artificial neural networks (ANN) trained in deep learning leave open questions with respect to at least two areas of human cognition. Firstly, being sub-symbolic, it is unclear how they relate to 'Jackendoff's challenges'[1] (Jackendoff, 2002, §3.5) and to higher-order, symbolic processing as observed, for instance, in sentence processing (Goucha et al., 2017; Frankland and Greene, 2020a). Secondly, despite being inspired by the human brain and potentially useful for neuro-scientific research (Yang and Wang, 2020), ANNs differ from biological neural networks. The first issue is addressed by Vector Symbolic Architectures (VSA; Gayler, 2003; Schlegel et al., 2022), which define symbolic operations on high-dimensional numerical vectors.

The second issue is addressed by biological architectures, where high-dimensional vectors receive a neural interpretation in terms of spiking patterns (Eliasmith, 2013). Formal semantics provides symbolic systems for analysing natural languages. However, as Lücking and Ginzburg (2023, p. 149) argue, it is questionable whether traditional, 'anti-representationalist' formal semantics, which assigns truth conditions directly to sentences (Bezuidenhout, 2006) also lends itself to cognitive interpretations.

This is different with a Type Theory with Records (TTR; Cooper, 2023), which even has a neural interpretation (Cooper, 2019a). Indeed there has been a wide range of work in this formalism, introduced in section 3, which includes the modelling of intensionality and mental attitudes (Cooper, 2005, 2023), quantified NPs (Cooper, 2013; Lücking and Ginzburg, 2022; Cooper, 2023),

[1] Namely 'The massiveness of the binding problem', 'The problem of 2', 'The problem of variables', and 'Binding in working memory vs. long-term memory'.

Proceedings of the 4th Natural Logic Meets Machine Learning Workshop (NALOMA23), pages 41–50
June 20, 2023.

co-predication and dot types in lexical innovation, frame semantics for temporal reasoning, reasoning in hypothetical contexts (Cooper, 2011), spatial reasoning (Dobnik and Cooper, 2017), enthymematic reasoning (Breitholtz, 2020), self- and other-repair (Purver, 2006; Ginzburg et al., 2014), negation (Cooper and Ginzburg, 2012), non-sentential utterance resolution (Fernández et al., 2007; Ginzburg, 2012), iconic gesture (Lücking, 2016), multimodality (Lücking and Ginzburg, 2023) and symbol grounding (Larsson, 2015, 2021).

Accordingly, this paper offers a first attempt to combine TTR with a biologically-based VSA, namely the Semantic Pointer Architecture (SPA) of Eliasmith (2013). Sections 2 and 3 provide a brief overview of semantic pointers and TTR, respectively. How to 'translate' TTR objects into SPA is addressed in Section 4. We conclude in Section 5.

2 SPA (and NEF)

> [...] semantic pointers are neural representations that are aptly described by by high-dimensional vectors, are generated by compressing more sophisticated representations, and can be used to access those more sophisticated representations through decompression [...]. (Eliasmith, 2013, p. 83)

Hence, there are three perspectives on or levels of description for semantic pointers, namely (i) in terms of neural activation, (ii) as (high-dimensional) vectors, and (iii) as symbols. In this paper, we will not be concerned with the neural level beyond the assumption that there are biologically plausible neural mechanisms underlying what happens on the levels of vectors and, most central to our concerns, the level of symbols. Here, we simply refer to and make use of the Neural Engineering Framework (Eliasmith and Anderson, 2003) and its Python implementation Nengo (Bekolay et al., 2014).

Schlegel et al. (2022) in their very useful survey of VSAs offer a comparison of different approaches in terms of four distinct parameters:

Hypervector selection: When selecting vectors to represent basic entities one aims to create maximally different encodings. Higher dimensional vector spaces offer sufficient space to maintain a large class of vectors distinct and moreover, they have the useful property that two random vectors are with very high probability quasi-orthogonal. A common strategy is to use a real range which is normally distributed with a mean of 0 and a variance of $1/D$ where D defines the number of dimensions.

Similarity measurement: VSAs use similarity metrics to evaluate vector representations, in particular, to assess whether the represented symbols have a related meaning. The similarity metric plays the essential role of selecting the correct denoised vector from the database and to ensure a robust operation of VSAs. The dot product of two vectors A, B is standardly computed as the sum of the product of their components, as in (1a). This is the basis for defining the cosine between two vectors as in (1b) in terms of the dot product and the vectors' lengths:

(1) a. $$A \cdot B = \sum_{k=0}^{D-1} a_k b_k$$

b. $$cos\ \theta = \frac{A \cdot B}{||A|| * ||B||}$$

Following most VSA approaches, we use cosine as a measure of similarity. Given (1b), this reduces to the dot product when the vectors are normalized (i.e., of length 1). If $A \cdot B \approx 1$, the vectors are (nearly) identical. For any vector A,

(2) $A \cdot A \approx 1$

Bundling: VSAs use a bundling operator to superimpose (or overlay) given hypervectors. Plate (1997) argues that a bundling operator must satisfy unstructured similarity preservation, namely $A + B$ is similar to A and to B and to any bundle $A + C$ that contains one of the vectors. Bundling is typically handled using vector addition, but in the approach adopted here this requires a normalization step to a vector length of one.

Binding: Binding $\times$ is used to connect two vectors, e.g., role-filler pairs. The output is again a vector from the same vector space. Plate (1997) argues that binding needs to satisfy:

- Non-similarity of bindees to output: $A \times B \not\approx A, B$
- Similarity preservation: $A \approx A'$, $B \approx B'$ implies $A \times B \approx A' \times B'$
- 'x' is invertible: if $C = A \times B$, there exists A^{-1} such that $C \times A^{-1} = B$

In the current paper we generally follow the approach known as Holographic Reduced Representations (HRR), first defined by Plate (1991), which is the approach utilized by Eliasmith and implemented in Nengo. However, as Eliasmith notes, one could make different choices if clear motivation for these arises. Specifically, with respect to binding we use circular convolution $C = A \circledast B$ defined as follows in a space of dimension D:

(3) **Circular convolution**

$$c_j = \sum_{k=0}^{D-1} b_k a_{j-k(modD)}$$

for $j \in \{0, \ldots, D-1\}$

Circular convolution approximates the standard tensor outer product by summing over all of its (wrap-around) diagonals. This operator is commutative as well as associative. Circular correlation provides an approximated inverse for circular convolution used for unbinding. The inverse is defined in (4a), exemplified in (4b), and its use for unbinding is given in (4c):[2]

(4) **Inverse for circular convolution**

a. $a_j^{-1} = a_{D-j(modD)}$
where $j \in \{0, \ldots, D-1\}$

b. In other words: $\langle a_0, a_1, \ldots, a_{D-1} \rangle^{-}1 = \langle a_0, a_{D-1}, \ldots, a_1 \rangle$

c. $A \circledast B \circledast B^{-1} \approx A$

In what follows, we use B' for B^{-1}.

3 TTR

We give a brief sketch of those aspects of TTR which we will use in this paper. For more detailed accounts see Cooper (2023).

$s : T$ represents a judgement that s is of type T. Types may be either *basic* or *complex* (in the sense that they are structured objects which have types or other objects introduced in the theory as components). One basic type that we will use is *Ind*, the type of individuals; another is *Real*, the type of real numbers.

[2]In algebra an element A's multiplicative inverse A^{-1} is by definition an element such that $A \times A^{-1} = 1$ (the unit element of multiplication). An approximate inverse of an element A ApproxInv(A)$^{-1}$ is one where $A \times ApproxInv(A)^{-1} \approx 1$.

Among the complex types are *ptypes* which are constructed from a predicate and arguments of appropriate types as specified for the predicate. Examples are 'man(a)', 'see(a,b)' where a, b : *Ind*. The objects or *witnesses* of ptypes can be thought of as situations, states or events in the world which instantiate the type. Thus s : man(a) can be glossed as "s is a situation which shows (or proves) that a is a man".

Another kind of complex type is *record types*. In TTR *records* are modelled as a labelled set consisting of a finite set of fields. Each field is an ordered pair, $\langle \ell, o \rangle$, where ℓ is a *label* (drawn from a countably infinite stock of labels) and o is an object which is a witness of some type. No two fields of a record can contain the same label. Importantly, o can itself be a record.

A *record type* is like a record except that the fields are of the form $\langle \ell, T \rangle$ where ℓ is a label as before and T is a type. The basic intuition is that a record, r is a witness for a record type, T, just in case for each field, $\langle \ell_i, T_i \rangle$, in T there is a field, $\langle \ell_i, o_i \rangle$, in r where $o_i : T_i$. (Note that this allows for the record to have additional fields with labels not included in the fields of the record type.)

The types within fields in record types may *depend* on objects which can be found in the record which is being tested as a witness for the record type. We use a graphical display to represent both records and record types where each line represents a field. Example (5) represents the type of records which can be used to model situations where a man runs.

$$(5) \quad \begin{bmatrix} \text{ref} & : & \textit{Ind} \\ \text{c}_{\text{man}} & : & \text{man(ref)} \\ \text{c}_{\text{run}} & : & \text{run(ref)} \end{bmatrix}$$

A record of this type would be of the form

$$(6) \quad \begin{bmatrix} \text{ref} & = & a \\ \text{c}_{\text{man}} & = & s \\ \text{c}_{\text{run}} & = & e \\ \ldots & & \end{bmatrix}$$

where a : *Ind*, s : man(a) and e : run(a).

Some of our types will contain *manifest fields* like the c_{man}-field below:

$$(7) \quad \begin{bmatrix} \text{ref} & : & \text{Ind} \\ \text{c}_{\text{man}}{=}s_{23} & : & \text{man(ref)} \end{bmatrix}$$

Here, $[c_{man}{=}s_{23}{:}\text{man(ref)}]$ is a convenient notation for $[c_{man}{:}\text{man(ref)}_{s_{23}}]$ where $\text{man(ref)}_{s_{23}}$ is a *singleton type*. If $a : T$, then T_a is a singleton type and $b : T_a$ iff $b = a$.[3] Manifest fields allow us to progressively specify what values are required for the fields in a type.

It is possible to combine record types. Suppose that we have two record types C_1 and C_2:

$$(8)\quad C_1 = \begin{bmatrix} \text{x} : \mathit{Ind} \\ c_{man} : \text{man(x)} \end{bmatrix}$$

$$C_2 = \begin{bmatrix} \text{x} : \mathit{Ind} \\ c_{run} : \text{run(x)} \end{bmatrix}$$

In this case, $C_1 \wedge C_2$ is a type; more specifically, a meet type. In general if T_1 and T_2 are types then $T_1 \wedge T_2$ is a type and $a : T_1 \wedge T_2$ iff $a : T_1$ and $a : T_2$. A meet type $T_1 \wedge T_2$ of two record types can be simplified to a new record type by a process similar to unification in feature-based systems. If T_1 and T_2 are record types then there will be a type $T_1 \dot{\wedge} T_2$ equivalent to $T_1 \wedge T_2$ (in the sense that something will be of the first type if and only if it is of the second type). The operation $\dot{\wedge}$ is referred to as merge.

$$(9)\quad C_1 \dot{\wedge} C_2 = \begin{bmatrix} \text{x} : \mathit{Ind} \\ c_{man} : \text{man(x)} \\ c_{run} : \text{run(x)} \end{bmatrix}$$

We will introduce further details of TTR as we need them in subsequent sections.

4 Relating SPA and TTR

4.1 The basic idea

We define a mapping, σ, from types in TTR to patterns (types) of neural activity represented as vectors in SPA[4]. On the basis of this we define neural judgement conditions of the form "agent A judges s to be of type T if a particular neural condition involving $\sigma(T)$ holds. The connective here is a conditional rather than a biconditional because we allow more than one pattern of neural activity to correspond to the same TTR judgement. For example, A may judge s to be of T because of, say visual perception, or because s has been stored in memory corresponding to the witness cache discussed in Cooper (2019b). This is in contrast to the proposal in Cooper (2019a) which defines a function from types to patterns (types) of neural activity but does not take the additional step of giving neural judgement conditions. The move from representing types to representing judgements, which belong to the theory of action defined on the theory of types, appears to us to be a conceptual improvement. Essentially, the correspondence we define characterizes the brain activity of an agent when engaged in an act of making a type judgement, rather than simply giving a neural representation of a type. This seems promising for building a theory of how an embodied agent perceives its environment rather than creating a neural representation of a type without specifying how it would link to the world.

Another way in which the approach taken here differs from that of Cooper (2019a) is that the approach to representing the structure of complex types relies on the vector operations used in SPA, such as circular convolution, rather than the phasing of neural activity as in Cooper (2019a) following in a tradition of neural modelling stemming from Shastri (1999). This raises a question of whether the modelling in terms of vector operations reveals enough structure which we will leave open in this paper.

Our aim in this paper is to begin mapping out a possible correspondence between TTR and SPA. We do not yet have a complete definition and there are a number of questions about what we have so far. Nevertheless, we hope that what we have represents a promising beginning. Below, we often use $T \sim \mathbf{T}$ to mean $\sigma(T) = \mathbf{T}$. We will also often use $\mathbf{T}$ to represent $\sigma(T)$

We will frequently let equality or near similarity between two patterns of neural activation in SPA terms characterize TTR neural judgement conditions. In doing this we will exploit the fact the the dot product of two (nearly) identical vectors $\mathbf{a}$ and $\mathbf{b}$, $\mathbf{a} \cdot \mathbf{b}$ is approximately equal to 1 (see Eliasmith, 2013, p. 389).

4.2 Basic types

We will use semantic pointers to correspond to basic TTR types. For basic types, we assume a

[3] Cooper (2023) uses a modification of this characterization of singleton types: if a is of some type, then T_a is a singleton type. $b : T_a$ iff $b : T$ and $b = a$. This allows for there to be types T_a where $a \not: T$. Such types have no witnesses.

[4] In this paper, we are not concerned with the converse mapping, from SPA to TTR.

function β that provides a unique semantic pointer corresponding to each basic type and that the function σ is defined relative to β:

(10) If T is a basic type, $\sigma_\beta(T) = \beta(T)$

We will suppress the β-subscript on σ in what follows.

4.3 Judgements

In TTR, judgements involving basic perceptual types can be made either using a classifier or based on a witness cache (Larsson, 2020). Type judgements based on classifiers take real-valued (e.g. perceptual) inputs.

In SPA, as exemplified by the MNIST dataset (Deng, 2012) and perceptual/cortical modelling, a classifier can be implemented as a hierarchical statistical model, which constructs representations of the input, which in turn are mapped into mechanistic SPA models (Tang and Eliasmith, 2010). At the highest level of the hierarchy, we have compressed representation summarising what has been presented to the lowest level. Following Eliasmith (2013), this compressed representation is a semantic pointer.

To judge whether a situation s is of a (perceptual) type T, the perception of s by an agent A generates a representation (in the form of neural activity, e.g. on V1, the primary visual cortex) $\mathbf{s_A}$ (A's take on s in the terminology of Larsson, 2020). A hierarchical statistical model, call it κ, when fed $\mathbf{s_A}$ as input to the lowest level of κ (e.g. V1) produces a compressed representation (neural activity) $\kappa[\mathbf{s_A}]$ on the highest level (IT, the inferotemporal cortex) of κ—see Figure 1 for an illustration. The semantic pointer $\mathbf{T}$ specifies a certain type of activity on the highest level of κ, and if this activity is triggered by A perceiving s, this corresponds to A judging s to be of type T. If T is a perceptual basic type related to the statistical model κ, then the neural judgement condition can be expressed as (11a) or equivalently (11b) .

(11) a. $s :_A T$ if $\kappa[\mathbf{s_A}] \approx \mathbf{T}$

b. $s :_A T$ if $\kappa[\mathbf{s_A}] \cdot \mathbf{T} \approx 1$

Below we will often suppress the A-subscript on ':'.

Type judgements can also be based on a witness cache. The witness cache in TTR is a function F

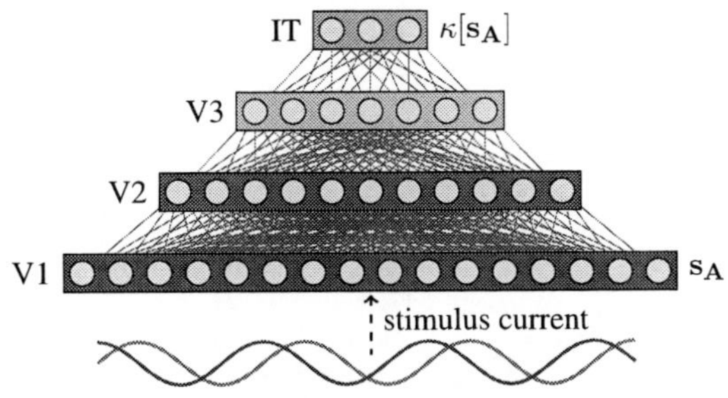

Figure 1: Illustration of hierarchical statistical model κ. To the left of each layer is the name of the layer, and to the right is the activity in that layer.

that takes a type T and returns a set of objects so that $x : T$ if $x \in F(T)$. We can let $\mathbf{F}$ be a structure that binds types with a bundling of semantic pointers $\mathbf{a_0} + \mathbf{a_1} + \ldots + a_n$, for example

(12) $\mathbf{F} = (\mathbf{Ind} \circledast (\mathbf{a} + \mathbf{b} + \ldots)) + (\mathbf{Int} \circledast (\mathbf{1} + \mathbf{2} + \ldots)) + \ldots$

In SPA, a bundle is similar to any of its elements. However, this similarity is more approximate than similarity between near-identical vectors. For this reason, we do not require the dot product of bundle and element to be 1, but only that it does not approximate 0:

(13) $(\mathbf{A_1} + \mathbf{A_2} + \ldots + \mathbf{A_n}) \cdot \mathbf{A_i} \not\approx 0$, $(1 \leq i \leq n)$

Given this, type checking can be done by looking up the witness cache in $\mathbf{F}$ and checking its similarity to the object:

(14) $x : T$ if $\mathbf{F} \circledast \mathbf{T}' \approxeq \mathbf{x}$

where we use $\approxeq$ so that this means

(15) $x : T$ if $\mathbf{F} \circledast \mathbf{T}' \cdot \mathbf{x} \not\approx 0$

(15) says that the vector which results from unbinding $\mathbf{T}$ associated with type T from $\mathbf{F}$ is (approximately) identical to the semantic pointer $\mathbf{a}$. For example:

(16) $a : \mathit{Ind}$ if $\mathbf{F} \circledast \mathbf{Ind}' \approxeq \mathbf{a}$

See Figure 2 for an example.[5]

[5]The code for this and the following examples can be found at `https://github.com/aluecking/SPA-TTR`.

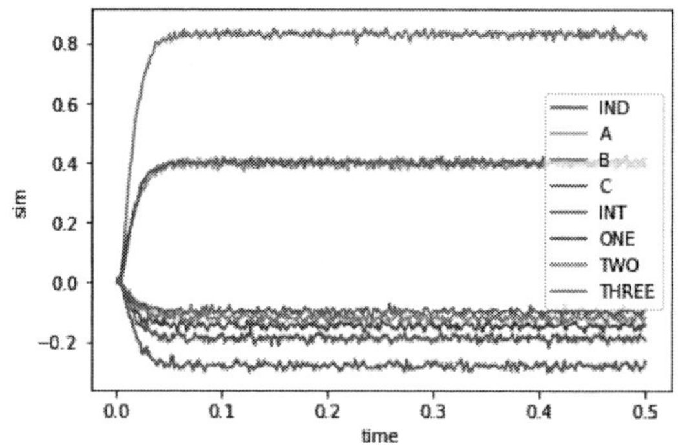

Figure 2: Given an **F** structure consisting of pointers for two basic types *IND* and *INT* bound to three object pointers each—*A*, *B*, *C*, respectively *ONE*, *TWO*, *THREE*—the (correct) result of unbinding **F** with $\mathbf{IND}'$ is approxiately ($\approxeq$) similar to pointers **A**, **B** and **C**.

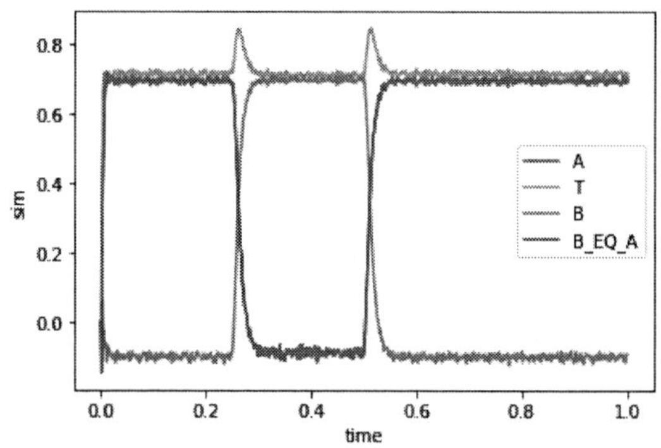

Figure 3: The similarity of T_a with b is only high if $b \approx a$. Comparing the similarity of $\mathbf{T} + \mathbf{a}$ ($t < 0.25\,\mathrm{s}$), $\mathbf{T} + \mathbf{b}$ ($0.25\,\mathrm{s} < t < 0.5\,\mathrm{s}$) and $\mathbf{T} + \mathbf{b} = \mathbf{a}$ (notated 'B_EQ_A'; $t > 0.5\,\mathrm{s}$) to all pointers in question (note that 'A' is masked by 'B_EQ_A').

4.4 Singleton types

A special case is typechecking for singleton types $T_a \sqsubseteq T$. We define the SPA structure to correspond to singleton types thus:

(17) $T_a \sim (\mathbf{T} + \mathbf{a})$

To check if $b : T_a$, we can check the equality $\mathbf{a} \approx \mathbf{b}$ and that $b : T$:

(18) $b : T_a$ if $\mathbf{a} \approx \mathbf{b}$ and $b : T$

—see Figure 3.

4.5 Labelled sets

Many structures in TTR are defined as labelled sets. We take labelled sets in TTR to correspond to SPA structures according to the following:

(19) $$\{\langle \ell_1, x_1\rangle, \ldots, \langle \ell_n, s_n\rangle\} \sim$$
$$\boldsymbol{\ell_1} \circledast \mathbf{d_1} + \ldots + \boldsymbol{\ell_n} \circledast \mathbf{x_n}$$

This move, however, involves treating labels as proper pointers, that is, compressed high(er) level semantic representations, which seems to be at odds with the status of labels as arbitrary bookkeeping devices. A potential way for reconciliation is to think of labels as indicating functional roles, as is initially attested in fMRI studies on processing, where it has been found that general agency (e.g., *owl-as-agent*) is represented in different cortical regions than narrow agency (e.g., *owl-as-chaser*) (Frankland and Greene, 2020b). This is reminiscent of an inferential view of thematic roles (Dowty, 1991), which seem to justify a semantic pointer representation, but poses the question whether this approach extends to all labels.

Labelled sets are sets of ordered pairs where the first item in each pair is a label. In SPA-TTR, we are using the binding operator $\circledast$ to associate two SPA terms. In both frameworks, given an item x and structure associating items (in TTR, a set S of ordered pairs of items; in SPA, a vector $\mathbf{S}$ as shown above) it is possible to retrieve the item y which x is associated with in S. In TTR, this is done by finding a pair $\langle x, t\rangle$ in S. In SPA-TTR, this is done by unbinding y from a binding $x \circledast y$ in S.

An important difference between TTR and SPA is that in TTR, it is easy to retrieve the labels that are used in a record type, which then enables relabelling the record as needed. In SPA-TTR, retrieving the labels requires probing $\mathbf{S}$ for the presence of each of a (finite) set of labels. If the set of labels is large, this may be inefficient. We do not offer a full solution to this problem here, but leave it for future work. However, we believe that a solution can be to keep around an index of the labels used in different record types.

4.6 Record types

We will not attempt here to represent TTR records in SPA, but focus instead of record types. Since TTR record types are labelled sets where the labels are paired with types, we use our SPA coding of labelled sets for record types.

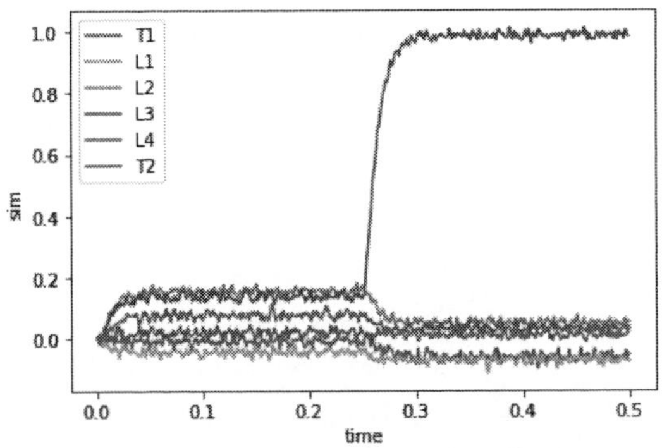

Figure 4: Recovering $\mathbf{T_2}$ from its path $\mathbf{T_1} \circledast \mathbf{L_1} \circledast \mathbf{L_2} \circledast \mathbf{L_3} \circledast \mathbf{L_4}$ is successful, but lossy as can be seen by comparison to querying $\mathbf{T_2}$ directly starting from 0.25 s.

(20) $$\begin{bmatrix} \ell_1 & : & T_1 \\ \ldots & & \\ \ell_n & : & T_n \end{bmatrix} \sim$$

$$\ell_1 \circledast \mathbf{T_1} + \ldots + \ell_\mathbf{n} \circledast \mathbf{T_n}$$

4.7 Paths in record types

In TTR, labels coinjoined by '.' form paths in records and record types. We can use unbinding in SPA to achieve something similar. If T_1 is a record type and T_2 is a type and $T_1.\ell_1.\ldots.\ell_m : T_2$ and $T_1 \sim \mathbf{P_1}, T_2 \sim \mathbf{P_2}, \ell_i \sim \mathbf{L_i}(1 \leq i \leq m)$ then

(21) $\mathbf{P_1} \circledast \mathbf{L_1'} \circledast \ldots \circledast \mathbf{L_m'} \approx \mathbf{P_2}$

We can recover **P2** (i.e., type T_2) from $\mathbf{P_1}$ by following the path $\mathbf{L_1'} \circledast \ldots \circledast \mathbf{L_m'}$, that is, by unbinding it with all the pointers used to construct it. Note that this retrieval is lossy, as illustrated in terms of a path consisting of four labels in Figure 4.

4.8 Meet and Merge

We take both the meet type $T_1 \wedge T_2$ of two types T_1 and T_2 and the merge $T_1 \dot{\wedge} T_2$ of two record types T_1 and T_2 to correspond to the SPA summing operation $+$.

(22) a. $T_1 \wedge T_2 \sim \mathbf{T_1} + \mathbf{T_2}$ for types T_1 and T_2

b. $T_1 \dot{\wedge} T_2 \sim \mathbf{T_1} + \mathbf{T_2}$ for record types T_1 and T_2

c. $\sigma(T_1 \wedge T_2) = \sigma(T_1 \dot{\wedge} T_2) = \mathbf{T_1} + \mathbf{T_2}$

The SPA summing operation is distributive in the same way that $\dot{\wedge}$ is—'binding distributes over bundling' (Schlegel et al., 2022, p. 4536)[6]—, so that

(23) $(\ell_\mathbf{1} \circledast \mathbf{T_1} + \ell_\mathbf{1} \circledast \mathbf{T_2} = (\ell_\mathbf{1} \circledast (\mathbf{T_1} + \mathbf{T_2}))$

corresponding to

(24) $[\ell_1{:}T_1] \dot{\wedge} [\ell_1{:}T_2] = [\ell_1{:}T_1 \dot{\wedge} T_2]$

Conflating $\wedge$ and $\dot{\wedge}$ means we are not making a distinction between $T_1 \wedge T_2$ and $T_1 \dot{\wedge} T_2$ for record types T_1, T_2 (for non-record types, they work in the same way also in TTR.).

4.9 Ptypes

Cooper (2023) defines a ptype $P(a_1, \ldots, a_n)$ as representing a labelled set $\{\langle \text{pred}, P \rangle, \langle \text{arg}_1, a_1 \rangle, \ldots \langle \text{arg}_1, a_n \rangle\}$. We follow this, so that e.g.

(25) a. run(a) $\sim$ (**pred** $\circledast$ **run** $+$ **arg1** $\circledast$ **a**)

b. hug(a,b) $\sim$ (**pred** $\circledast$ **hug** $+$ **arg1** $\circledast$ **a** $+$ **arg2** $\circledast$ **b**)

An important area for future research is to enable classifier-based judgements of sensory input as being of ptypes and record types involving ptypes. For example, given a situation s where a boy hugs a dog, we want an agent A's take on s to be judged to be of a complex type involving properties and relations.

4.10 Subtyping

Since subtyping can be defined in terms of a TTR equality between two types, this could appear to be a means of formulating the corresponding SPA-TTR definition:

(26) a. $T_1 \sqsubseteq T_2$ if $T_1 \dot{\wedge} T_2 = T_1 \sim (\mathbf{T_1} + \mathbf{T_2}) \approx \mathbf{T_1}$

b. $\sigma(T_1 \sqsubseteq T_2) = (\mathbf{T_1} + \mathbf{T_2}) \approx \mathbf{T_1}$

For example,

[6] In fact, in Nengo the vocabulary parses of, e.g., 'A * B + A * C' and 'A * (B + C)' result in the same vector.

(27) $\sigma(\begin{bmatrix}\text{x:a}\\\text{y:b}\end{bmatrix} \sqsubseteq \begin{bmatrix}\text{x:a}\end{bmatrix}) =$
$((\mathbf{x}\circledast\mathbf{a}+\mathbf{y}\circledast\mathbf{b})+(\mathbf{x}\circledast\mathbf{a})) \approx (\mathbf{x}\circledast\mathbf{a}+\mathbf{y}\circledast\mathbf{b})$

However, the above solution does not work because (27) holds only if $\mathbf{T_1} = \mathbf{T_2}$, which is of course a much stronger requirement than subtyping. An alternative could be to apply an element-wise maximum function:

(28) a. $T_1 \sqsubseteq T_2$ iff $T_1 \dot{\wedge} T_2 = T_1 \sim \max(\mathbf{T_1}, \mathbf{T_2}) \approx \mathbf{T_1}$

b. $\sigma(T_1 \sqsubseteq T_2) = \max(\mathbf{T_1}, \mathbf{T_2}) \approx \mathbf{T_1}$

The similarity of the maximum is indeed larger than the (cosine) similarity of supertype and subtype (see `https://github.com/aluecking/SPA-TTR`). However, further work is needed to further specify and verify this proposal.

4.11 Functions

TTR functions can be represented as labelled sets, but doing so says nothing about how they are applied to arguments. For this reason, we will here be focusing on TTR functions as lambda abstracted expressions. We will not offer a complete account of TTR functions in SPA here, but only offer some initial remarks.

For instance, assume we have a function

(29) $\lambda r{:}\begin{bmatrix}\text{x : Ind}\end{bmatrix} \cdot \begin{bmatrix}\text{c} & : & \text{run}(r.\text{x})\end{bmatrix}$

This function corresponds to the following mini-network:

(30)

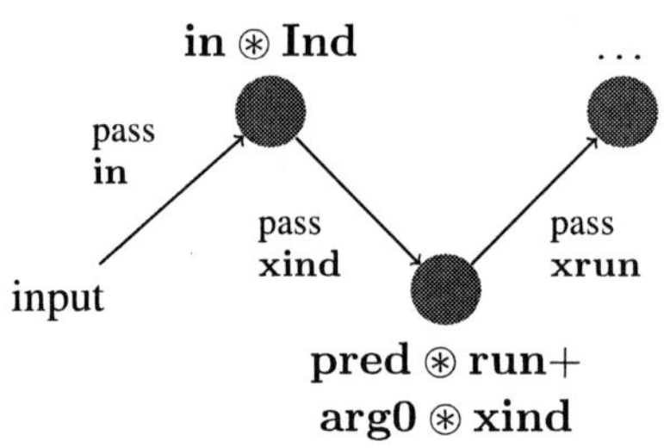

Or in SPA syntax:

```
d = 64 # use vectors of 64 dimensions
xind = spa.State(vocab=d)
xrun = spa.State(vocab=d)

input * spa.sym("IND") >> xind
xind * spa.sym("ARG0") + spa.sym("PRED *
    RUN")) >> xrun
```

where 'input' can, for instance, receive activation from another network such as κ (see (11b)) or sequentially range over (any subset of) the objects bound to **IND** in the witness cache (see (12)):

```
def inputs(t):
    if t < 0.25:
        return "A"
    elif t < 0.5:
        return "B"
    ...

input = spa.Transcode(inputs,
    output_vocab=d)
```

5 Summary and conclusions

In this paper, we took initial steps towards relating TTR to SPA, with mostly encouraging results. We accounted for basic types, perceptual and cache-based judgements, singleton types, record types, meet types and merging of record types, ptypes, and subtyping. As indicated above, more work is needed to account for subtyping and judgements involving ptypes. Work is ongoing to cover more aspects of TTR in SPA, including records and functions. In addition to these, several TTR elements remain to be covered, including join types, asymmetric merge, and type stratification to name but a few.

The benefit of succeeding with this effort would be a true hybrid between formal and neural semantics that could potentially have the benefits of both but the drawbacks of neither. We also hope that this work may throw light on many puzzling issues regarding the relation between formal and neural semantics.

Acknowledgements

The work of the first and second author was supported by a grant from the Swedish Research Council (VR project 2014-39) for the establishment of the Centre for Linguistic Theory and Studies in Probability (CLASP) at the University of Gothenburg.

References

Trevor Bekolay, James Bergstra, Eric Hunsberger, Travis DeWolf, Terrence Stewart, Daniel Rasmussen, Xuan Choo, Aaron Voelker, and Chris Eliasmith. 2014. Nengo: a Python tool for building large-scale functional brain models. Frontiers in Neuroinformatics, 7.

Anne Bezuidenhout. 2006. VP-ellipsis and the case for representationalism in semantics. ProtoSociology,

22:140–168. Compositionality, Concepts and Representations II.

Ellen Breitholtz. 2020. Enthymemes and Topoi in Dialogue: the use of common sense reasoning in conversation. Brill.

Robin Cooper. 2005. Austinian truth, attitudes and type theory. Research on Language and Computation, 3(4):333–362.

Robin Cooper. 2011. Copredication, quantification and frames. In Logical Aspects of Computational Linguistics (LACL 2011). Springer.

Robin Cooper. 2013. Clarification and generalized quantifiers. Dialogue and Discourse, 4:1–25.

Robin Cooper. 2019a. Representing types as neural events. Journal of Logic, Language and Information, 28:131–155.

Robin Cooper. 2019b. Types as Learnable Cognitive Resources in PyTTR. In Cleo Condoravdi and Tracy Holloway King, editors, Tokens of Meaning: Papers in Honor of Lauri Karttunen. CSLI Publications.

Robin Cooper. 2023. From Perception to Communication. Number 16 in Oxford Studies in Semantics and Pragmatics. Oxford University Press, Oxford, UK.

Robin Cooper and Jonathan Ginzburg. 2012. Negative inquisitiveness and alternatives-based negation. In Logic, Language and Meaning, pages 32–41. Springer.

Li Deng. 2012. The MNIST database of handwritten digit images for machine learning research. IEEE Signal Processing Magazine, 29(6):141–142.

Simon Dobnik and Robin Cooper. 2017. Interfacing language, spatial perception and cognition in type theory with records. Journal of Language Modelling, 5(2):273–301.

David Dowty. 1991. Thematic proto-roles and argument selection. Language, 67(3):547–619.

Chris Eliasmith. 2013. How to Build a Brain: A Neural Architecture for Biological Cognition. Oxford University Press, Oxford.

Chris Eliasmith and Charles H. Anderson. 2003. Neural Engineering. Computational Neuroscience. MIT Press, Cambridge, MA.

Raquel Fernández, Jonathan Ginzburg, and Shalom Lappin. 2007. Classifying ellipsis in dialogue: A machine learning approach. Computational Linguistics, 33(3):397–427.

Steven M. Frankland and Joshua D. Greene. 2020a. Concepts and compositionality: In search of the brain's language of thought. Annual Review of Psychology, 71(1):273–303. PMID: 31550985.

Steven M. Frankland and Joshua D. Greene. 2020b. Two ways to build a thought: Distinct forms of compositional semantic representation across brain regions. Cerebral Cortex, 30(6):3838–3855.

Ross W. Gayler. 2003. Vector symbolic architectures answer Jackendoff's challenges for cognitive neuroscience. pages 133–138. Slightly updated 2004 in paper cs/0412059 on arXiv.

Jonathan Ginzburg. 2012. The interactive stance. Oxford University Press.

Jonathan Ginzburg, Raquel Fernández, and David Schlangen. 2014. Disfluencies as intra-utterance dialogue moves. Semantics and Pragmatics, 7(9):1–64.

Tomás Goucha, Emiliano Zaccarella, and Angela D. Friederici. 2017. A revival of *Homo loquens* as a builder of labeled structures: Neurocognitive considerations. Neuroscience & Biobehavioral Reviews, 81:213–224. The Biology of Language.

Ray Jackendoff. 2002. Foundations of Language. Oxford University Press, Oxford, UK.

Staffan Larsson. 2015. Formal semantics for perceptual classification. Journal of Logic and Computation, 25(2):335–369. Published online 2013-12-18.

Staffan Larsson. 2020. Discrete and probabilistic classifier-based semantics. In Proceedings of the Probability and Meaning Conference (PaM 2020), pages 62–68.

Staffan Larsson. 2021. The role of definitions in coordinating on perceptual meanings. In Proceedings of the Workshop on the Semantics and Pragmatics of Dialogue (SemDial 2021).

Andy Lücking. 2016. Modeling co-verbal gesture perception in type theory with records. In Proceedings of the 2016 Federated Conference on Computer Science and Information Systems, pages 383–392.

Andy Lücking and Jonathan Ginzburg. 2022. Referential transparency as the proper treatment for quantification. Semantics and Pragmatics, 15(4):1–56.

Andy Lücking and Jonathan Ginzburg. 2023. Leading voices: Dialogue semantics, cognitive science, and the polyphonic structure of multimodal interaction. Language and Cognition, 15(1):148–172.

Tony Plate. 1991. Holographic reduced representations: Convolution algebra for compositional distributed representations. In Proceedings of the 12th International Joint Conference on Artificial Intelligence, IJCAI'91, pages 30–35.

Tony Plate. 1997. A common framework for distributed representation schemes for compositional structure. Connectionist systems for knowledge representation and deduction, pages 15–34.

M. Purver. 2006. CLARIE: Handling clarification requests in a dialogue system. Research on Language & Computation, 4(2):259–288.

Kenny Schlegel, Peer Neubert, and Peter Protzel. 2022. A comparison of vector symbolic architectures. Artificial Intelligence Review, 55(6):4523–4555.

Lokendra Shastri. 1999. Advances in SHRUTI – a neurally motivated model of relational knowledge representation and rapid inference using temporal synchrony. Applied Intelligence, 11(1):79–108.

Yichuan Tang and Chris Eliasmith. 2010. Deep networks for robust visual recognition. In Proceedings of the 27 th International Conference on Machine Learning, pages 1055–1062.

Guangyu Robert Yang and Xiao-Jing Wang. 2020. Artificial neural networks for neuroscientists: A primer. Neuron, 107(6):1048–1070.

Triadic temporal representations and deformations

Tim Fernando
School of Computer Science & Statistics
Trinity College Dublin, Ireland
Tim.Fernando@tcd.ie

Abstract

Triadic representations that temporally order events and states are described, consisting of strings and sets of strings of bounded but refinable granularities. The strings are compressed according to J.A. Wheeler's dictum *it from bit*, with bits given by statives and non-statives alike. A choice of vocabulary and constraints expressed in that vocabulary shape representations of cause-and-effect with deformations characteristic, Mumford posits, of patterns at various levels of cognitive processing. These deformations point to an ongoing process of learning, formulated as grammatical inference of finite automata, structured around Goguen and Burstall's institutions.

1 Introduction

What does a string $\mathfrak{s}$ that is assigned a probability by a language model describe? Over a range of uses, $\mathfrak{s}$ is uttered at time S to describe an event occurrence at time E. Reichenbach (1947) suggests that S is connected to E by a *reference time* R, traversing three corners

language (S), agent (R), and world (E)

of a triangle that is arguably congruent with the well-known symbol-thought-referent triangle from Ogden and Richards (1923), page 11. Reichenbach derives nine *fundamental forms*, including the *simple past* (1) and *present perfect* (2), by positioning R relative to S and to E (with $<$ as "earlier than").

$R < S$ and $R = E$	(Ed ate.)	(1)
$S = R$ and $E < R$	(Ed has eaten.)	(2)

For fundamental forms, S, R and E may be considered points; but for extended tenses with the present participle (*-ing*) and temporal adverbs (such as *yesterday*), E and R are stretched to temporal intervals. E and R have since been refined in various ways (e.g., Moens and Steedman, 1988; Kamp and Reyle, 1993; Nelken and Francez, 1995; Asher and Lascarides, 2003; Klein, 2009; Kehler, 2022), and the speech time S extended to an interval timing an utterance event so that

	language	agent	world
Ogden and Richards, 1923	symbol	thought	referent
Reichenbach, 1947	S	R	E
Liang and Potts, 2015	u	s	d
Goguen and Burstall, 1992	*Sen*	**Sig**	**Mod**

Table 1: Some triads

(†) the meaning $[\![\mathfrak{s}]\!]$ of a simple declarative sentence $\mathfrak{s}$ is a relation $u\ [\![\mathfrak{s}]\!]e$ between an utterance u (with time S) and a described situation e (with time E).

(†) is an early formulation of a *relation theory of meaning* (Barwise and Perry, 1983, page 19) that is developed further in, for example, Cooper and Ginzburg (2015); Cooper (2023). Left out of (†) is the reference time R which Reichenbach uses as a bridge between S and E. That is, (†) is dyadic, supplying an utterance u and denotation $d = e$ for a *linguistic object* $\langle u, s, d\rangle$ in Liang and Potts (2015) with*out* a semantic representation s which Table 1 aligns with R (under *agent* and *thought*, sandwiched between the utterance u of the string $\mathfrak{s}$, and the denotation[1] d).

Among the semantic representations considered in Liang and Potts (2015) are "distributed representations — vectors and matrices" that feed into string probabilities from language models. The present work focuses on semantic representations that support probabilities via more familiar logical forms. These forms describe patterns of events that are linked below to *pattern theory* (Grenander and Miller, 2007), which D. Mumford defines as

> the analysis of the patterns generated by the world in any modality, with all their

[1] (†)'s denotation e is closer to the "conventional meaning" than to the "communicative intent" discussed in Bender and Koller (2020)'s critique of large neural language models.

Proceedings of the 4th Natural Logic Meets Machine Learning Workshop (NALOMA23), pages 51–61
June 20, 2023.

> naturally occurring complexity and ambiguity, with the goal of reconstructing the processes, objects and events that produced them. [Mumford, 1994, page 187]

Expecting such a link flies against Mumford's view that pattern theory "stands in opposition to the accepted analysis of thought in terms of logic" but is less surprising if indeed "pattern theory contains the germs of a universal theory of thought itself" [page 221]. Fundamental to pattern theory is a "principle of realism" stating that

> the pattern should not merely describe the 'pure' situation that underlies reality but the 'deformed' situation that is actually observed in which the pure pattern may be hard to recognize. This generalizes, for example, Chomsky's idea of the deep structure of an utterance vs. its surface structure, where deep $\sim$ pure and surface $\sim$ deformed. [Mumford, 2019, page 203]

Mumford (1994) claims "the world does not have an infinite repertoire of different tricks which it uses to disguise what is going on" and picks out four types of deformations "encountered at all levels of cognitive processing." These deformations can be seen in cause-and-effect representations formed below which share two basic features with the information-theoretic formulation of pattern theory in Mumford (1994)

(i) a finite space Ω of functions f from a finite set of variables to a finite set of values, and

(ii) an encoding of f such that code(f) has a length which is minimized to reconstruct the world w that f is about.

By restricting to finite sets, (i) bounds the granularity of the representation, imposing a finite precision on values. The blurring here is an instance of one of Mumford's four types of deformations, taken up in section 3 below, where it is associated with a move from records to record types (Cooper and Ginzburg, 2015; Cooper, 2023). The code lengths mentioned in (ii) are used in Mumford (1994) for an approach to Bayesian maximum likelihood estimates based on Shannon's optimal coding theorem. The function f and the world w it is about in (ii) can, from the perspective of Table 1 above, be likened to an utterance u and denotation d that u is about. Even for simple declarative sentences $\mathfrak{s}$, however, the leap from an utterance u of $\mathfrak{s}$ to its denotation d is an enormous one, inviting the question: would a mediating representation s between u and d not help? Arguably, such a representation s is what code(f) in (ii) is, although it is not obvious from Mumford (1994) or Grenander and Miller (2007) what form s might take for an utterance u of a declarative sentence.

The semantic representations s below describe not only events such as denotations d but also utterances u of pieces $\mathfrak{s}$ of language ranging from multi-sentential discourses (as in Kamp and Reyle (1993)) down to subsentential units. Following (ii), code lengths are minimized in section 2, but appealing in this case to Wheeler (1990)'s dictum *it from bit*. To illustrate the idea, consider Reichenbach's simple past (1) and present perfect (2), reformulated as strings $\boxed{E,R}\boxed{S}$ and $\boxed{E}\boxed{R,S}$ respectively, both of length 2, and the past perfect (3) represented by the string $\boxed{E}\boxed{R}\boxed{S}$ of length 3.[2]

$$R < S \text{ and } E < R \qquad \text{(Ed had eaten.)} \qquad (3)$$

If we focus on S and E and throw R out, we can compress all three strings to $\boxed{E}\boxed{S}$ representing the relation $E < S$ common to (1), (2) and (3), saying u is about an event d in u's past. The details are given in section 2, where strings are formulated as models of predicate logic (e.g., Libkin, 2004) with a specified signature fixing granularity. While that granularity is bounded by finite sets in (i), ever larger finite signatures Σ can be collected in a category **Sig** that a functor **Mod** maps contravariantly to sets of Σ-models and a functor *Sen* maps covariantly to sets of Σ-sentences. The triad *Sen*, **Sig**, **Mod** occupies the bottom row of Table 1, and can be organised into a logical system called an *institution* (Goguen and Burstall, 1992; Goguen, 2006). An amalgamation property enjoyed by well-behaved *algebraic* institutions (e.g., Sannella and Tarlecki, 2015) is, however, damaged by compression. This is explained in section 3, where compression is equated with another of Mumford (1994)'s deformations, domain warping. Further deformations are noted that shape the sample space Ω on which a probability measure is defined (yielding probabilities that are front and center in pattern theory). What makes the strings here interesting is

[2] $\boxed{E,R}\boxed{S}$ is written E, R-S in Reichenbach (1947). Using boxes instead of curly braces {, } for sets qua string symbols suggests reading a comic strip (e.g., Fernando, 2015).

	active	stative
Kleene, 1956	input cell	inner cell
dynamic logic	program	proposition
action language	elementary action	fluent
sig (A, V)	act $\in A$	variable $\in dom(V)$

Table 2: Deconstructing a transition $q \xrightarrow{a} q'$

that they represent some of "processes, objects and events" that produce patterns. These patterns include certain causes and effects, packaged as *event nuclei* in Moens and Steedman (1988), that can be framed around transitions in finite automata amenable to probabilistic elaboration.

2 Strings as compressed models

The neural nets for which Kleene (1956) introduced finite automata have cells of two kinds: *input cells* which could either fire or not, and *inner cells* which could take one of finitely many values, depending on the input cells and inner cells that feed into them. For neural nets with k input cells $\mathcal{N}_1, \ldots, \mathcal{N}_k$, Kleene forms an alphabet of 2^k symbols (one for each subset of $\{\mathcal{N}_1, \ldots, \mathcal{N}_k\}$), and from m inner cells $\mathcal{M}_1, \ldots, \mathcal{M}_m$, generates m-tuples $(v_1, \ldots, v_m)$ consisting of values v_i that $\mathcal{M}_i$ can take. A couple of notational conventions will prove handy below. For any integer $j > 0$, let us write $[j]$ for the set of j integers from 1 to j

$$[j] := \{1, 2, \ldots, j\}.$$

Next, given a set-valued function V, let $\prod V$ be the set of V-records, where a V-*record* is a function r with the same domain as V that maps each $\mathbf{x}$ in $dom(V)$ to an element $r(\mathbf{x})$ of $V(\mathbf{x})$. It is often convenient to write $\prod V$ out as $\prod_{\mathbf{x} \in dom(V)} V(\mathbf{x})$. For example, if each inner cell $\mathcal{M}_i$ can take s_i many values, then the set

$$\prod_{i \in [m]} [s_i] \;\cong\; [s_1] \times \cdots \times [s_m]$$

of functions r mapping $i \in [m]$ to one of s_i many values, $r(i)$, is isomorphic to the set of m-tuples $(v_1, \ldots, v_m)$ assigning inner cell $\mathcal{M}_i$ the v_i^{th} of s_i values. $\prod_{i \in [m]} [s_i]$ can serve as the set of states between which any set $a \subseteq \{\mathcal{N}_1, \ldots, \mathcal{N}_k\}$ of input cells can label a binary relation $\xrightarrow{a}$ of transitions $q \xrightarrow{a} q'$ from state q to q'. Table 2 aligns inner cells with the stative sides q, q' of $q \xrightarrow{a} q'$, and input cells with the active middle a. The stative/active dichotomy is perhaps most famously developed in the proposition/program distinction drawn in *Dynamic Logic* (Harel et al., 2000), but the conception of a transition label a as a set of firing input cells puts us on a different course.

Input cells become elementary actions in *action languages* (Gelfond and Lifschitz, 1998), where a transition label a (called an action) is a set of elementary actions, while a state q is described by values taken by certain fluents[3] corresponding to inner cells. See the penultimate row of Table 2. The bottom row Table 2 brings out what Kleene (1956) and Gelfond and Lifschitz (1998) have in common through the following rudimentary notion of *sig*nature.

Definition. A *sig* is a pair (A, V) consisting of a finite set A of *acts* and a function V with a finite domain, $dom(V)$, of *variables* $\mathbf{x}$, each paired with a finite set $V(\mathbf{x})$ of *values* that $\mathbf{x}$ can take.

A sig (A, V) provides a finite vocabulary of acts in A and statives (given by variables and values) in V. Statives are central to works such as Dowty (1979), where they are the basis of an aspectual calculus.

An instructive example is provided by the leap below from (4) to (5) by virtue of the entailment (6) from *bought* to *owns* proposed in Hosseini (2020); see also Hosseini et al. (2019).

Facebook bought Instagram (4)

Facebook owns Instagram (5)

$$\boxed{\text{bought}(x, y)} \Rightarrow \boxed{\text{owns}(x, y)} \qquad (6)$$

(6) assumes no change in ownership of y after x bought y; this assumption may fail depending on subsequent events. (7) repairs this flaw in (6) by applying the operator BECOME to the (untensed) stative own(x, y) to produce a non-stative BECOME(own(x, y)).

$$\boxed{\text{buy}(x, y)} \Rightarrow \boxed{\text{BECOME}(\text{own}(x, y))} \qquad (7)$$

The meaning of BECOME in (7) is brought out in a transition (8) labelled by buy(x, y) from a state where x does *not* own y to a state where x does (0 marking falsity, and 1 truth).

$$\boxed{(\text{own}(x, y), 0)} \xrightarrow{\text{buy}(x,y)} \boxed{(\text{own}(x, y), 1)} \qquad (8)$$

[3] Action languages belong to the symbolic AI tradition blazed by John McCarthy, who adopted Newton's term *fluent* for a state variable (the value of which may change over time).

Entailments such as (9), however, make clear there is more to buy(x, y) than BECOME(own(x, y)).

$$\boxed{\text{buy}(x,y)} \Rightarrow \boxed{\text{pay}(x,y)} \quad (9)$$

It is easy enough to replace buy(x, y) in (8) by BECOME(own(x, y)), but the question is: can we reduce (4) to a transition $q \xrightarrow{a} q'$ without leaving out some of the details, such as *Facebook paid for Instagram*, implicit in (4) according to (9)? There are two directions along which to extend $q \xrightarrow{a} q'$. First, more than one act may go into the transition label a on the understanding that

(‡) $q \xrightarrow{a} q'$ says: the acts in a execute concurrently to move from q to q'.

Second, we may break $q \xrightarrow{a} q'$ down to a chain (10) of n transitions $q_{i-1} \xrightarrow{a_i} q_i$ between states q_{i-1} and q_i labelled by sets a_i of acts from $q_0 = q$ to $q_n = q'$.

$$q_0 \xrightarrow{a_1} q_1 \xrightarrow{a_2} q_2 \cdots \xrightarrow{a_n} q_n. \quad (10)$$

Now, for any fixed n in (10), is it not conceivable that some (if not every) transition $q_{i-1} \xrightarrow{a_i} q_i$ can be refined to a longer transition chain from q_{i-1} to q_i? Perhaps so. But if we use a sig (A, V) to require of an (A, V)-chain (10) that

$$q_0, q_i \text{ are } V\text{-records} \quad \text{and} \quad a_i \subseteq A \quad (11)$$

then it is more plausible that further refinements of (10) would involve stepping from the sig (A, V) to a suitably larger sig (A', V'). Just what suitably larger means, we take up in the next section. In the meantime, note that the transition (8) serves as an account of buy(x, y) for the sig (A, V), where A is $\{$buy$(x, y)\}$ and the function V is, as a set of ordered pairs $(\mathbf{x}, V(\mathbf{x}))$, the singleton

$$V = \{(\text{own}(x,y), \{0,1\})\}$$

with exactly one variable own(x, y), the values of which are either 0 or 1.

Next, fixing a sig (A, V), let us package the (A, V)-chain (10) as the $(A.V)$-string

$$(q_0, a_1)(q_1, a_2) \cdots (q_{n-1}, a_n)(q_n, a_{n+1})$$

of $n + 1$ pairs (q_{i-1}, a_i), where $a_{n+1} = \emptyset$. To simplify notation, let us assume

(NAP) **no act in A is an ordered pair**

so that given a pair (q, a) of $q \in \prod V$ and $a \subseteq A$, we can recover from the union $\alpha = q \cup a$, the label a and state q

$$a = \alpha \cap A \quad \text{and} \quad q = \alpha \setminus A$$

through set complementation

$$X \setminus Y \ := \ \{x \in X \mid x \notin Y\}.$$

Flattening (q, a) to $q \cup a$, the transition (8) becomes the string

$$\boxed{(\text{own}(x,y),0),\, \text{buy}(x,y)}\boxed{(\text{own}(x,y),1)}$$

of length 2, the first symbol/box of which has stative part $\boxed{(\text{own}(x,y),0)}$ and active part $\boxed{\text{buy}(x,y)}$. The partiality of a sig suggests widening the range of (A, V)-strings beyond those obtained from transition chains (10) that (11) ties to a sig (A, V). We drop (11) to accommodate larger sigs (A', V') where q_0, q_i are V'-records and $a \subseteq A'$. In particular, we might extract the $(A, \emptyset)$-string $a_1 a_2 \cdots a_n$ from (10) where each of the labels a_i is a subset of A. Adding the restriction that each a_i be non-empty leads us to Durand and Schwer (2008), where an *S-word* is defined to be a string of non-empty sets. But why exclude the empty box $\square$ from an S-word?

Under (‡), it is natural to assume the value of a variable cannot change without an act, leading to the following principle of inertia

$$\text{whenever } q \xrightarrow{\square} q', \ q = q'. \quad (12)$$

The transition $q \xrightarrow{\square} q$ hardly describes any change, and arguably carries zero information, suggesting that any occurrence of the empty box $\square$ in $a_1 \cdots a_n$ be deleted. Similarly, if we extract the $(\emptyset, V)$-string $q_0 q_1 \cdots q_n$ of states from (10) for a sig (A, V) where each q_i in (10) is a V-record. then any stutter qq might be deleted from $q_0 q_1 \cdots q_n$, as in the *block compression* $bc(s)$ *of* a string s

$$bc(s) \ := \ \begin{cases} s & \text{if length}(s) \leq 1 \\ bc(qs') & \text{if } s = qqs' \\ q\, bc(q's') & \text{if } s = qq's' \\ & \quad \text{where } q \neq q' \end{cases}$$

(Fernando, 2015).

For sigs (A, V) where neither A nor V need be empty, let us collect the (A, V)-*boxes* from which we form (A, V)-strings in the alphabet

$$\mathcal{B}_{A,V} \ := \{a \cup r \mid a \subseteq A \text{ and } r \in \prod V\}$$

and define the *A-compression* $\kappa_A(s)$ *of* a string $s \in \mathcal{B}_{A,V}{}^*$ by induction on the length of s

$$\kappa_A(s) := \begin{cases} \epsilon & \text{if } s = \epsilon \text{ or } s = \Box \\ s & \text{else if length}(s) = 1 \end{cases}$$

and for strings of length ≥ 2,

$$\kappa_A(\alpha\alpha' s) := \begin{cases} \kappa_A(\alpha' s) & \text{if } \alpha = \Box \text{ or} \\ & \alpha = \alpha' \setminus A \\ \alpha\, \kappa_A(\alpha' s) & \text{otherwise} \end{cases}$$

(Fernando, 2022). Clearly,

$$\kappa_A(\kappa_A(s)) = \kappa_A(s)$$

and in case A or V is empty,

$$\kappa_A(s) = \begin{cases} s \text{ without } \Box\text{'s} & \text{if } V = \emptyset \\ bc(s) & \text{else if } A = \emptyset. \end{cases}$$

A-compression κ_A implements the Aristotelian dictum *no time without change* (e.g., Coope, 2001), or better:

$$\text{no time}_{A,V} \text{ without change}_{A,V}. \qquad (13)$$

To see this, it is useful to construe a non-empty string of (A, V)-boxes as a *model of Monadic Second-Order* logic (MSO), with MSO-sentences that capture the sets of such strings accepted by finite automata via a satisfaction relation $\models$ (e.g., Libkin, 2004, Theorem 7.21).

More precisely, let us collect the possible input/output pairs of V-records in the set

$$\sum V := \{(\mathbf{x}, c) \mid \mathbf{x} \in dom(V) \text{ and } c \in V(\mathbf{x})\}.$$

Let the *vocabulary* of a sig (A, V) be the union

$$voc(A, V) := A \cup \sum V$$

of A with $\sum V$, and for every $u \in voc(A, V)$, let us form a fresh unary relation symbol P_u. P_u is interpreted relative to a string $\alpha_1 \cdots \alpha_n \in \mathcal{B}^+_{A,V}$ of (A, V)-boxes α_i as the set

$$[\![P_u]\!]_{\alpha_1 \cdots \alpha_n} := \{i \in [n] \mid u \in \alpha_i\}$$

of string positions i where u occurs. Hence, the disjunction $\bigvee_{u \in A} P_u(i)$ says: some act from A occurs at i. In addition to unary relation symbols P_u, there is a binary relation S that is interpreted as the successor (+1) relation

$$[\![S]\!]_{\alpha_1 \cdots \alpha_n} := \{(i, i+1) \mid i \in [n-1]\}$$

on string positions. We can conjoin $P_u(i)$ with the negation of the claim that u occurs at a successor of i for the formula

$$\delta_u(i) := P_u(i) \wedge \neg\exists j(iSj \wedge P_u(j))$$

which for $u \in \sum V$ can be paraphrased: u holds at position i but not immediately afterwards. Accordingly, the MSO-sentence

$$ntwoc_{A,V} := \forall i \,(\bigvee_{u \in A} P_u(i) \vee \bigvee_{u \in \sum V} \delta_u(i))$$

says:

> at every string position, some act from A occurs or some V-stative holds but not immediately afterwards

which amounts to (13), assuming string positions represent time$_{A,V}$, and change$_{A,V}$ is communicated through the set

$$\{P_u \mid u \in A\} \cup \{\delta_u \mid u \in \sum V\} \qquad (14)$$

of active predicates P_u ($u \in A$) and stative changes δ_u ($u \in \sum V$). It turns out $ntwoc_{A,V}$ expresses the effect of A-compressing strings over the alphabet $\mathcal{B}_{A,V}$.

Theorem. *For all* $s \in \mathcal{B}_{A,V}{}^+$,

$$s \models ntwoc_{A,V} \iff s = \kappa_A(s).$$

The theorem is proved by a routine induction on the length of s. Our account of patterns based on (A, V) will center around the set

$$\mathbf{Mod}(A, V) := \{\kappa_A(s) \mid s \in \mathcal{B}_{A,V}{}^+\}$$

of (A, V)-*models*, which the theorem above equates with (A, V)-strings satisfying $ntwoc_{A,V}$

$$\mathbf{Mod}(A, V) = \{s \in \mathcal{B}_{A,V}{}^+ \mid s \models ntwoc_{A,V}\}.$$

There are two kinds of variables here against which to apply Quine (1950)'s prescription that

> to be assumed as an entity is to be assumed as a value of a variable (p.228)

— viz., so-called variables $\mathbf{x}$ in $dom(V)$ with values in $V(\mathbf{x})$, and variables such as i, j that occur free and bound in $\delta_u(i)$, and range over time. The latter time variables link $ntwoc_{A,V}$ to J.A. Wheeler's dictum *it from bit*

> every it — every particle, every field of force, even the space**time** continuum itself — derives its function, its meaning, its very existence entirely — even if in some contexts indirectly — from the apparatus-elicited answers to yes-or-no questions, binary choices, bits. [Wheeler, 1990, p. 5]

The string positions $[n]$ of $\alpha_1 \cdots \alpha_n$ are constrained by $ntwoc_{A,V}$ to changes$_{A,V}$ (14) observed through the apparatus MSO$_{A,V}$. The power of that apparatus is bound by the sig (A, V) which is refined in the next section to expand what can be observed, uncovering deformations along the way.

3 Projections and deformations

Relaxing the finiteness assumptions built into a sig, let us fix a pair $(\mathsf{Act}, \mathsf{Val})$ of

(a) a set Act of acts, none of which is an ordered pair (building in the no-act-pair assumption (NAP) from the previous section), and

(b) a function Val from variables $\mathbf{x}$ to sets $\mathsf{Val}(\mathbf{x})$ of values that $\mathbf{x}$ can take.

A *finite blurring of* Val is a function V whose domain, $dom(V)$, is a finite subset of $dom(\mathsf{Val})$ such that for each $\mathbf{x} \in dom(V)$, $V(\mathbf{x})$ is a finite partition of $\mathsf{Val}(\mathbf{x})$. Thus, $\sum V$ is finite even if $\sum \mathsf{Val}$ is not (due to $dom(\mathsf{Val})$ or some $\mathbf{x} \in dom(\mathsf{Val})$ with infinite $\mathsf{Val}(\mathbf{x})$). The intuition is that V approximates Val up to finite precision.[4] Under *it-from-bit*, the finite approximations V have an arguably stronger claim to reality than the idealization Val.

With this in mind, let us define an $(\mathsf{Act}, \mathsf{Val})$-*sig* to be a pair (A, V) of a finite subset A of Act and a finite blurring V of Val. $(\mathsf{Act}, \mathsf{Val})$-sigs can be partially ordered as follows. (A, V) *is refined by* (A', V'), written $(A, V) \preceq (A', V')$, if $A \subseteq A'$, $dom(V) \subseteq dom(V')$ and for each $\mathbf{x} \in dom(V)$, the partition $V'(\mathbf{x})$ refines $V(\mathbf{x})$ in the usual sense (i.e., every value-set from $V'(\mathbf{x})$ is a subset of some value-set from $V(\mathbf{x})$). Assuming $(A, V) \preceq (A', V')$, let

(a) the (A, V)-*reduct of* an (A', V')-box α' be the (A, V)-box

$$\rho_{A,V}(\alpha') := (\alpha' \cap A) \cup \alpha'_V$$

where $\alpha'_V \in \prod V$ maps $\mathbf{x} \in dom(V)$ to the unique $V(\mathbf{x})$-equivalence class that includes the value-set that α' assigns to $\mathbf{x}$[5]

(b) the (A, V)-*reduct of* a string of (A', V')-boxes be its componentwise (A, V)-reduct

$$\rho_{A,V}(\alpha'_1 \cdots \alpha'_n) := \rho_{A,V}(\alpha'_1) \cdots \rho_{A,V}(\alpha'_n)$$

(c) the (A, V)-*projection of* an (A', V')-model s' be the A-compression of its (A, V)-reduct

$$\kappa_{A,V}(s') := \kappa_A(\rho_{A,V}(s')) .$$

For example, given an (A', V')-model $\alpha'_1 \cdots \alpha'_n$, its $(\emptyset, V)$-projection for $V \neq \emptyset$ is the block compression

$$bc((\alpha'_1)_V \cdots (\alpha'_n)_V)$$

and its $(A, \emptyset)$-projection is the S-word

$$(\alpha'_1 \cap A) \cdots (\alpha'_n \cap A) \text{ without } \square\text{'s.}$$

Returning to Reichenbach's fundamental forms, if we treat the points E, R, S as acts, then $\boxed{E}\boxed{S}$ is the $(\{E, S\}, \emptyset)$-projection of each of the strings

$$\boxed{E, R}\boxed{S}, \quad \boxed{E}\boxed{R, S}, \quad \boxed{E}\boxed{R}\boxed{S}$$

for the simple past (1), present perfect (2) and past perfect (3), respectively. Shortening $\boxed{E}\boxed{R}\boxed{S}$ to $\boxed{E}\boxed{S}$ is an instance of *domain warping* (Mumford, 1994, p. 196) inasmuch as the domain of a string, as an MSO-model, is its set of string positions. In general, any change in string length from s' to $\kappa_{A,V}(s')$ can be put down to the compression κ_A built into $\kappa_{A,V}$. A-compression is required by the Theorem from the previous section if an (A, V)-model is to satisfy $ntwoc_{A,V}$. It is also indispensible for representing finite subsets of the real line $\mathbb{R}$ (a popular model of time) as strings of finite length — e.g., $\{0, 1, e, \pi\} \subseteq \mathbb{R}$ as $\boxed{0}\boxed{1}\boxed{e}\boxed{\pi}$ depicting $0 < 1 < e < \pi$. Clearly, $\mathbb{R}$ can be reconstructed by a projective (inverse) limit over string representations of its finite subsets. Take away A-compression and we lose this reconstruction.

Unfortunately, A-compression complicates the amalgamation of different (A, V)-projections.

[4] The reduction of $dom(\mathsf{Val})$ to a finite subset $dom(V)$ is compatible with the usual restriction on records to finitely many fields; the blurring of values in $\mathsf{Val}(\mathbf{x})$ to subsets of $\mathsf{Val}(x)$ in $V(\mathbf{x})$ suggests a further move to record types (Cooper and Ginzburg, 2015; Cooper, 2023).

[5] When V is V' restricted to $dom(V)$ (i.e., $V \subseteq V'$), the (A, V)-reduct $\rho_{A,V}(\alpha')$ of α' is just the intersection $\alpha' \cap voc(A, V)$.

This can be seen by looking once more at Reichenbach (1947)'s fundamental forms. Inasmuch as $R < S$ can be pictured as $\boxed{R}\boxed{S}$ and $R = E$ as $\boxed{E,R}$, the step from the conjunction (1) of $R < S$ and $R = E$ to the string $\boxed{E,R}\boxed{S}$ can be expressed as

$$\boxed{R}\boxed{S} \,\&\, \boxed{E,R} = \boxed{E,R}\boxed{S}.$$

On the other hand, the conjunction of $R < S$ and $R < E$ for the *posterior past* yields three different strings

$$\boxed{R}\boxed{E,S},\ \boxed{R}\boxed{E}\boxed{S},\ \boxed{R}\boxed{S}\boxed{E}, \qquad (15)$$

each of which has $(\{R,S\},\emptyset)$-projection $\boxed{R}\boxed{S}$ for $R < S$ and $(\{R,E\},\emptyset)$-projection $\boxed{R}\boxed{E}$ for $R < E$. (Similarly, for the *anterior future* from $S < R$ ad $E < R$). The non-uniqueness here can be summarized as

($*$) the presheaf **Mod** does not satisfy the gluing condition necessary for a sheaf

which we presently unpack. **Mod** is a *presheaf* insofar as **Mod** can be understood as a set-valued contravariant functor from the category **Sig** of (A,V)-sigs with morphisms given by the ordered pairs $((A,V),(A',V'))$ from refinement $\preceq$, where $\mathbf{Mod}((A',V'),(A,V))$ maps an (A',V')-model s' to its (A,V)-projection $\kappa_{A,V}(s')$

$$\text{i.e., } \mathbf{Mod}((A',V'),(A,V))(s') = \kappa_{A,V}(s').$$

Next, let us call two $(\mathsf{Act},\mathsf{Val})$-sigs (A_1,V_1) and (A_2,V_2) *compatible* if V_1 and V_2 agree on the intersection of their domains

$$\text{i.e., } (\forall \mathbf{x} \in dom(V_1) \cap dom(V_2))\ V_1(\mathbf{x}) = V_2(\mathbf{x})$$

making $(A_1 \cup A_2, V_1 \cup V_2)$ an $(\mathsf{Act},\mathsf{Val})$-sig. Given compatible sigs (A_1,V_1) ad (A_2,V_2), and (A_i,V_i)-models s_i for $i \in [2]$, let $s_1\&s_2$ be the set of all $(A_1 \cup A_2, V_1 \cup V_2)$-models s that project to s_1 and to s_2

$$\kappa_{A_1,V_1}(s) = s_1 \text{ and } \kappa_{A_2,V_2}(s) = s_2.$$

The gluing condition in ($*$) requires that the set $s_1\&s_2$ be a singleton whenever s_1 and s_2 agree on the $(\mathsf{Act},\mathsf{Val})$-sig $(A_1 \cap A_2, V_1 \cap V_2)$

$$\kappa_{A_1\cap A_2, V_1\cap V_2}(s_1) = \kappa_{A_1\cap A_2, V_1\cap V_2}(s_2).$$

This requirement is not met by $\boxed{R}\boxed{S}\&\boxed{R}\boxed{E}$, which consists of the three strings in (15).[6] Only the first string $\boxed{R}\boxed{E,S}$ would remain were we to drop A-compression from (A,V)-projection.[7]

Keeping A-compression, we shall give an account of the conjunction $s_1\&s_2$ above through a functor *Sen* from the category **Sig** mapping an $(\mathsf{Act},\mathsf{Val})$-sig (A,V) covariantly to a set $Sen(A,V)$ of (A,V)*-sentences*. There are as many choices of $Sen(A,V)$ as there are ways of defining the languages accepted by finite automata, the crucial requirement on $Sen(A,V)$ being that there be a relation $\models_{A,V}$ between (A,V)-models and (A,V)-sentences such that

(i) for every (A,V)-sentence φ, there is a finite automaton accepting the set

$$\mathbf{Mod}_{A,V}(\varphi) := \{s \in \mathbf{Mod}(A,V) \mid s \models_{A,V} \varphi\}$$

of (A,V)-models that satisfy φ (under $\models_{A,V}$)

and conversely,

(ii) for every subset L of $\mathbf{Mod}(A,V)$ that is accepted by some finite automaton, there is some (A,V)-sentence φ capturing L

$$L = \mathbf{Mod}_{A,V}(\varphi).$$

For concreteness, we may equate $Sen(A,V)$ with the set of $\mathrm{MSO}_{voc(A,V)}$-sentences. Now, whenever $(A,V) \preceq (A',V')$, let $Sen((A,V),(A',V'))$ map an (A,V)-sentence φ to an (A',V')-sentence $\langle (A,V),(A',V')\rangle\varphi$ such that

($**$) $\mathbf{Mod}_{A',V'}(\langle (A,V),(A',V')\rangle\varphi)$ is the set

$$\{s' \in \mathbf{Mod}(A',V') \mid \kappa_{A,V}(s') \models_{A,V} \varphi\}$$

of (A',V')-models whose (A,V)-projections satisfy φ.

($**$) is the *Satisfaction condition* characteristic of an *institution* (Goguen and Burstall, 1992). The existence of an (A',V')-sentence $\langle (A,V),(A',V')\rangle\varphi$ validating ($**$) follows from the regularity assumptions (i) and (ii) above, and

[6]In this case, $A_1 = \{R,S\}, A_2 = \{R,E\}, V_1 = V_2 = \emptyset$. In general, $(A,V) \preceq (A',V')$ implies $A \subseteq A'$ but not necessarily $V \subseteq V'$. To sidetep notational complications, however, our discussion of gluing will proceed with the simple case of $V \subseteq V'$.

[7]Gluing is referred to as *amalgamation* in, for example, Sannella and Tarlecki (2015), where it is admitted by algebraic institutions with reducts as projections.

the closure of regular languages under inverse images of relations such as $\kappa_{A,V}$ computed by finite-state transducers. Under ($**$), $\langle(A, V), (A', V')\rangle$ is a modal operator for $\kappa_{A,V}$, albeit not one of the primitive propositional connectives or quantifiers in MSO. Now, given two compatible sigs (A_1, V_1) and (A_2, V_2) and two (A_i, V_i)-sentences φ_i for $i \in [2]$, let us attach the modal operator $\langle(A_i, V_i), (A_1 \cup A_2, V_1 \cup V_2)\rangle$ to φ_i for

$$\psi_i := \langle(A_i, V_i), (A_1 \cup A_2, V_1 \cup V_2)\rangle\varphi_i$$

and observe that the conjunction $\psi_1 \wedge \psi_2$ captures $s_1 \& s_2$ provided φ_i captures s_i for $i \in [2]$. Such a conjunction is an instance of *multi-scale superposition* (Mumford, 1994, p. 195), the third of four types of deformations instantiated above (alongside blur, $\preceq$, and domain warping, $\kappa_{A,V}$).

The fourth of Mumford's deformations arises when examining cause-and-effect within a sig (A, V). For a handle on how acts $u \in A$ affect statives $v \in \sum V$, let us fix a function af with domain Act mapping every act $u \in \mathsf{Act}$ to a set $\mathsf{af}(u) \subseteq dom(\mathsf{Val})$ of variables that u can *af*fect. Given a pair $(\mathbf{x}, c) \in \sum V$, let us collect the acts in A that can affect $\mathbf{x}$ in

$$A_{(\mathbf{x},c)} := \{u \in A \mid \mathbf{x} \in \mathsf{af}(u)\}.$$

Next, we form an MSO-formula $\delta_v(i, j)$ saying v holds at i but not at its successor j

$$\delta_v(i, j) := iSj \wedge P_v(i) \wedge \neg P_v(j).$$

Building on our understanding (‡) of transitions $q \xrightarrow{a} q'$ and inertia (12) from section 2, let us agree that an (A, V)-model s is af-*inertial* if for every pair $v \in \sum V$,

$$\forall i \forall j\ (\delta_v(i, j) \supset \bigvee_{u \in A_v} P_u(i)) \qquad (16)$$

which is to say: any v-change in s occurs with an act in A that can affect v. One of the challenges in meeting (16) is that the act that affects v need not be in the finite subset A of Act. Indeed, an af-inertial string s may, for some $A_\circ \subset A$, have $(A_\circ, V)$-projection $\kappa_{A_\circ,V_\circ}(s)$ that is *not* af-inertial because (16) requires an act $u \in A \backslash A_\circ$ outside $A_\circ$. (A, V)-models s which are not af-inertial are "incompelete observations" called "interruptions" in Mumford (1994), page 196, that invite an expansion $A' \supseteq A$ of A and a search for af-inertial (A', V)-models s' that are *dense paraphrases* (Ye et al., 2022) of s insofar as $\kappa_{A,V}(s') = s$. The trigger (16) for refining sigs can be extended to more elaborate constraints such as

$$\forall i \forall j\ (\delta_v(i, j) \supset \bigvee_{u \in A_v} (P_u(i) \wedge \chi^u(i, j))) \qquad (17)$$

which conjoins $P_u(i)$ with a suitable description $\chi^u(i, j)$ of an *event nucleus* around the *culmination* u with a *preparatory process* at i and *consequent state* at j (Moens and Steedman, 1988). (17) reduces to (16) if χ^u is a tautology, but may otherwise take us outside (A, V), depending on how the preparatory process and consequent state are fleshed out. To keep the direction from state change to acts in (16), we can recast (17) as

$$\forall i \forall j\ ((P_u(i) \wedge iSj)\ \supset \chi^u(i, j)) \qquad (18)$$

for the reverse direction from acts to state change (and between (16) and (18), a cleaner interplay between A and V than in (17)).

For a concrete illustration, consider again

Facebook bought Instagram (4)

Facebook owns Instagram (5)

$\boxed{\text{bought}(x, y)} \Rightarrow \boxed{\text{owns}(x, y)}$ (6)

The step from (4) to (5) suggested by the tensed predicates in (6) becomes more inviting if we insert *has* before *bought* in (4), and less so with *had*.

Facebook bought Instagram. $\boxed{E, R \mid S}$ (19)

Facebk has bought Instagram. $\boxed{E \mid R, S}$ (20)

Facebk had bought Instagram. $\boxed{E \mid R \mid S}$ (21)

Without R, the strings in (19) to (21) collapse to

$\boxed{E \mid S}$ *Past(buy(facebook,instagram))*

The issue for (5) is: does the result *own(facebook, instagram)* of the *buy(facebook,instagram)*-event at E hold at the same box as S (assuming a sufficiently coarse notion of speech time so that S can serve both (4) and (5)). If *own(facebook,instagram)* coincides with R, the leap to (5) becomes easier from (20), if not from (19) or less, from (21). Expanding the sig (A, V), perhaps through (17), provides the ingredients for a more intricate account.

4 Conclusion

A triadic system $(\mathbf{Sig}, \mathbf{Mod}, \mathit{Sen})$ of finite-state representations is presented above, describing events and states through a vocabulary $(\mathsf{Act}, \mathsf{Val})$ of active and stative predicates. Finite fragments (A, V) of $(\mathsf{Act}, \mathsf{Val})$ are collected in **Sig**, from which **Mod** compresses strings of (A, V)-boxes, and *Sen* forms (A, V)-sentences defining sets of strings accepted by finite automata. As the compression on (A, V)-models can be computed by finite-state transducers, the (A, V)-sentences are closed under modal operators that turn the triad $(\mathbf{Sig}, \mathbf{Mod}, Sen)$ into an institution. Four types of deformations that Mumford claims shape patterns at various levels of cognitive processing can be discerned in these semantic representations

(D1) blur in approximating $(\mathsf{Act}, \mathsf{Val})$ by (A, V)

(D2) domain warping from compressing strings for (A, V)-models of $it_{A,V}$-*from*-$bit_{A,V}$

(D3) superposition implemented over (A, V)-sentences representing sets of (A, V)-models

(D4) interruptions marked by (A, V, af)-accounts of inertia and cause-and-effect.

The deformations point to the brittleness of the semantic representations: (D1) to the limited detail in any sig (A, V); (D2) to the dependence of a model's domain (i.e., time) on its vocabulary; (D3) to the need to step from an (A, V)-model to an (A, V)-sentence; and (D4) to the step from an (A, V)-sentence to a range of (A', V')-sentences over various refinements (A', V') of (A, V).

The steps here are roughly comparable to Pearl's "ladder of causation" with rungs for observing, doing, and imagining (Pearl and Mackenzie, 2018). To say more, the obvious next step would be to bring in probabilities and noise. That anything at all could be said before taking that step reflects the extent to which causal graphs can be drawn and paths in them found without numbers (in line with Pearl (2009)'s *Causal-Statistical Dichotomy*).

Staying with what is presented above, let us return to the question with which we began: what does a string $\mathfrak{s}$ that is assigned a probability by a language model describe? We have focused on the case where $\mathfrak{s}$ is uttered to describe a particular event or situation, ignoring examples such as (22) that are not restricted to any particular situations, or (23) that are just one of many opinions.

Facebook spreads lies. (22)

Facebook is evil. (23)

To support a range of situations and views, increasingly complex structures are proposed above around an explicit notion of granularity, signature. A signature provides a handle on the variation supported, to keep matters from getting out of hand. Try as we might to get things right, however, the concluding lines in Reichenbach (1947) are telling.

> The history of language shows that logical categories were not clearly seen in the beginnings of language but were the results of long developments; we therefore should not be astonished if actual language does not always fit the schema which we try to construct in symbolic logic. A mathematical language can be coordinated to actual language only in the sense of an approximation.

Computational linguists have long complained about the brittleness of semantic representations; it is time for semanticists to own it. Our representations are brittle because, as approximations, they get bits wrong. But mistakes (which experience/data corrects) feed learning, which is what grammatical inference and pattern theory are about, not to mention the engine behind the astonishing technological strides of recent years. By comparison, the approximations Reichenbach refers to are corrected at a glacial, ponderous pace. Though that too is learning. The main thrust of the present paper is to show how deformations from pattern theory drive us to steps up in abstractness — from a finite vocabulary to an expansion of it, around which strings and their projections are (contravariantly) formed, and further up to sets of strings and their (covariant) refinements. Nor can we stop at any fixed institution, except for constraints of space and time that force these complications to be taken up elsewhere.

And so, while it may be difficult to pin down what, in general, a string assigned a probability by a language model describes, this much can be said. The string is about an open process, approximated (as far as we can tell) by representations of bounded but refinable granularity. Fleshing this out, the technicalities above represent an attempt to marry (if you will) the information-theoretic approach to pattern theory outlined in Mumford (1994) with institutions, understood according to Goguen (2006) as an elaboration of C.S. Peirce's triadic theory of signs, semiotics (and perhaps, process of signing, semiosis; e.g., Atkin, 2023).

References

N. Asher and A. Lascarides. 2003. *Logics of Conversation*. Cambridge University Press.

A. Atkin. 2023. Peirce's theory of signs. In E.N. Zalta and U. Nodelman, editors, *The Stanford Encyclopedia of Philosophy*. Stanford, `https://plato.stanford.edu/archives/spr2023/entries/peirce-semiotics/`.

J. Barwise and J. Perry. 1983. *Situations and Attitudes*. MIT Press.

E.M. Bender and A. Koller. 2020. Climbing towards NLU: on meaning, form, and understanding in the age of data. In *Proceedings of the 58th Annual Meeting of the Association for Computational Linguistics*, pages 5185 – 5198.

U. Coope. 2001. Why does Aristotle say that there is no time without change? *Proceedings of the Aristotelian Society*, 101:359–367.

R. Cooper. 2023. *From Perception to Communication: A Theory of Types for Action and Meaning*. Oxford University Press.

R. Cooper and J. Ginzburg. 2015. TTR for natural language semantics. In S. Lappin and C. Fox, editors, *Handbook of Contemporary Semantic Theory*, second edition, pages 375–407. Wiley-Blackwell.

D.R. Dowty. 1979. *Word Meaning and Montague Grammar*. Reidel.

I.A. Durand and S.R. Schwer. 2008. A tool for reasoning about qualitative temporal information: the theory of S-languages with a Lisp implementation. *J. Univers. Comput. Sci.*, 14(20):3282–3306.

T. Fernando. 2015. The semantics of tense and aspect: a finite-state perspective. In S. Lappin and C. Fox, editors, *Handbook of Contemporary Semantic Theory*, second edition, pages 203–236. Wiley-Blackwell.

T. Fernando. 2022. Strings from neurons to language. In *Proc. Natural Logic meets Machine Learning III*, pages 1–10. ACL Anthology.

M. Gelfond and V. Lifschitz. 1998. Action languages. *Linköping Electronic Articles in Computer and Information Science*, 3(16).

J.A. Goguen. 2006. Institutions and Peirce's triadic semiotics. `https://cseweb.ucsd.edu/~goguen/projs/inst-sidebar.html`. Accessed: 2023-04-21.

J.A. Goguen and R.M. Burstall. 1992. Institutions: Abstract model theory for specification and programming. *Journal of the ACM*, 39(1):95–146.

U. Grenander and M. Miller. 2007. *Pattern Theory: From Representation to Inference*. Oxford University Press.

D. Harel, D. Kozen, and J. Tiuryn. 2000. *Dynamic Logic*. MIT Press.

M.J. Hosseini. 2020. Unsupervised learning of relational entailment graphs from text. PhD Thesis, School of Informatics, University of Edinburgh.

M.J. Hosseini, S.B. Cohen, M. Johnson, and M. Steedman. 2019. Duality of link prediction and entailment graph induction. In *Proc 57th ACL*, pages 4736–4746. Florence, Italy.

H. Kamp and U. Reyle. 1993. *From Discourse to Logic*. Kluwer Academic Publishers.

A. Kehler. 2022. Coherence establishment as a source of explanation in linguistic theory. *Annual Review of Linguistics*, 8:123–42.

S.C. Kleene. 1956. Representation of events in nerve nets and finite automata. In C. Shannon and J. McCarthy, editors, *Automata Studies*, pages 3–41. Princeton University Press.

W. Klein. 2009. How time is encoded. In *The Expression of Time*, pages 39–82. De Gruyter Mouton.

P. Liang and C. Potts. 2015. Bringing machine learning and compositional semantics together. *Annual Review of Linguistics*, 1:355–376.

L. Libkin. 2004. *Elements of Finite Model Theory*. Springer.

M. Moens and M. Steedman. 1988. Temporal ontology and temporal reference. *Computational Linguistics*, 14(2):15–28.

D. Mumford. 1994. Pattern Theory: a unifying perspective. In *First European Congress of Mathematics*, pages 187–224. Birkhäuser.

D. Mumford. 2019. A tribute to Ulf Grenander. *Quarterly of Applied Mathematics*, 77(2):201–206.

R. Nelken and N. Francez. 1995. Splitting the reference time: Temporal anaphora and quantification in DRT. In *Seventh Conference of the European Chapter of the Association for Computational Linguistics*.

C. K. Ogden and I. A. Richards. 1923. *The Meaning of Meaning: A Study of the Influence of Language upon Thought and of the Science of Symbolism*. Harcourt Brace Jovanovich, Inc.

J. Pearl. 2009. *Causality*, second edition. Cambridge University Press.

J. Pearl and D. Mackenzie. 2018. *The Book of Why*. Penguin.

W.V.O. Quine. 1950. *Methods of Logic*. Henry Holt and Co.

H. Reichenbach. 1947. The tenses of verbs. In *Elements of Symbolic Logic*, pages 287–298. Macmillan & Co.

D. Sannella and A. Tarlecki. 2015. The foundational legacy of ASL. In *Software, Services and Systems: Essays Dedicated to Martin Wirsing on the Occasion of His Retirement from the Chair of Programming and Software Engineering*, volume 8950 of LNCS, pages 253–272. Springer.

J.A. Wheeler. 1990. Information, physics, quantum: The search for links. In W. Zurek, editor, *Complexity, Entropy and the Physics of Information*, pages 3–28. Addison-Wesley.

B. Ye, J. Tu, E. Jezek, and J. Pustejovsky. 2022. Interpreting logical metonymy through dense paraphrasing. In *Proc 44th Annual Conference of the Cognitive Science Society*, pages 3541–3548.

Discourse Representation Structure Parsing for Chinese

Chunliu Wang, Xiao Zhang, Johan Bos
CLCG, University of Groningen
{chunliu.wang, xiao.zhang, johan.bos}@rug.nl

Abstract

Previous work has predominantly focused on monolingual English semantic parsing. We, instead, explore the feasibility of Chinese semantic parsing in the absence of labeled data for Chinese meaning representations. We describe the pipeline of automatically collecting the linearized Chinese meaning representation data for sequential-to-sequential neural networks. We further propose a test suite designed explicitly for Chinese semantic parsing, which provides fine-grained evaluation for parsing performance, where we aim to study Chinese parsing difficulties. Our experimental results show that the difficulty of Chinese semantic parsing is mainly caused by adverbs. Realizing Chinese parsing through machine translation and an English parser yields slightly lower performance than training a model directly on Chinese data.

1 Introduction

Semantic parsing is the task of transducing natural language text into semantic representations, which are expressed in logical forms underlying various grammar formalisms, such as abstract meaning representations (AMR, Wang et al. 2020; Bevilacqua et al. 2021), minimal recursion semantics (MRS, Horvat et al. 2015), and Discourse Representation Theory (DRT, Kamp and Reyle 1993). In this work, we explore the feasibility of parsing Chinese text to semantic representation based on Discourse Representation Structures (DRSs, Bos 2015a; van Noord et al. 2018), which are meaning representations proposed from DRT, a recursive first-order logic representation comprising of discourse referents (the entities introduced in the discourse) and relations between them.

Several neural parsers for DRS have been recently developed (Fancellu et al., 2019; Evang, 2019; van Noord et al., 2019; Liu et al., 2019; Wang et al., 2021; van Noord et al., 2020a) and reached remarkable performance, but mostly focused on monolingual English or some language using the Latin alphabet. Meaning representations are considered to be language-neutral, and texts with the same semantics but in different languages have the same meaning representation. The literature presents several examples of parsing multilingual text by training on monolingual English semantic representations (Ribeiro et al., 2021).

For the reason of relatively limited amounts of labeled gold-standard multilingual meaning representation data, multilingual text parsing relies on the source of silver English meaning representation data. As long as the meanings are expressed in a language-neutral way, this is a valid approach. However, named entities aren't usually, because they can (a) have different orthography for different languages using the same alphabet (in particular for location names, e.g., Berlin, Berlijn, Berlino, Berlynas) or (b) be written with a completely different character set, as is the case for Chinese.

Figure 1 shows a (nearly) language-neutral meaning representation for a simple English sentence. For non-English Latin alphabet languages, the named entities in the text are usually consistent with English, and the meaning in the form of a graph structure of the corresponding Discourse Representation (Discourse Representation Graph, DRG) would be identical to these languages (Bos, 2021), as shown in Figure 1. However, it would be rather absurd to expect a semantic parser for Chinese to produce meaning representations (with interlingual WordNet synsets) where proper names are anchored using the Latin alphabet using English (or any other language for that matter) orthography. We need to keep this important aspect in mind when evaluating semantic parsers for languages other than English.

However, for non-Latin alphabet languages, such as the widely used language of Chinese, is it feasible to use English meaning representation as the meaning representation of Chinese? Our objective

Proceedings of the 4th Natural Logic Meets Machine Learning Workshop (NALOMA23), pages 62–74
June 20, 2023.

is to investigate whether Chinese semantic parsing can achieve the same performance as English semantic parsing while using the same amount of data. We try to investigate whether it is necessary to develop a dedicated parser for Chinese, or whether it is possible to achieve a similar performance using an English parser by leveraging machine translation (MT) on Chinese. We provide inexpensively acquired silver-standard Chinese DRS data to implement our exploration: (1) We collect Chinese and English aligned texts from the Parallel Meaning Bank (PMB, Abzianidze et al. 2017), which provides parallel multilingual corpora including corresponding English meaning representation expressed in DRSs. (2) We leverage GIZA++ (Och and Ney, 2003) to align the word-segmented Chinese and English to obtain Chinese-English named entity alignment pairs, the resulting named entities are used to replace the named entities in our English semantic representation. (3) We train two monolingual parsers on the two languages separately, and then provide a set of fine-grained evaluation metrics to make better comparison between parsers. We aim to answer the following questions:

1. Can existing DRS parsing models achieve good results for Chinese? (RQ1)
2. What are the difficulties in semantic parsing for Chinese? (RQ2)
3. Is it feasible to use machine translation and an English parser to parse Chinese? How is it different from designing a special parser for Chinese? (RQ3)
4. How to conduct more fine-grained evaluation of experimental results and reduce the workload of manual evaluation? (RQ4)

2 Background

2.1 Discourse Representation Structure

DRS, as a kind of formal meaning representation, can be used to represent the semantic meaning of sentences and discourse. For the wide coverage of linguistic phenomena at quantification, negation, reference resolution, comparatives, discourse relations, and presupposition, DRT and DRS possess stronger semantic representation power than AMR. A DRS comprises discourse referents and conditions. However, some variants of DRS formats have been introduced in recent years, the format we employ throughout our work being one of them. We use a simplified DRS, which can be called Discourse Representation Graph (DRG) or Simplified Box Notation (SBN; Bos 2021). It discards explicit discourse references and variables while maintaining the same expressive power, as shown in Figure 2.

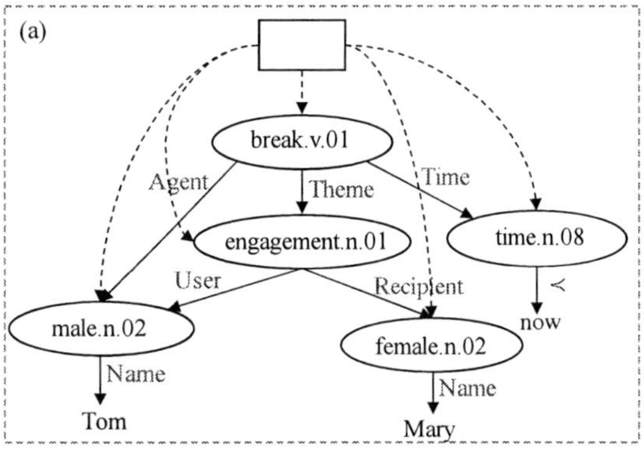

(b) male.n.02 Name "Tom"
break.v.01 Agent -1 Time +1 Theme +3
time.n.08 TPR now
male.n.02 EQU -3
engagement.n.01 User -1 Recipient +1
female.n.02 Name "Mary"

(c) EN: Tom broke off his engagement to Mary.
DE: Tom löste seine Verlobung mit Mary auf.
ZH: 汤姆与玛丽解除了婚约。

Figure 1: DRS in (a) graph format, (b) sequential box notation and (c) corresponding multilingual texts for English, German and Chinese.

As introduced by Bos (2021), DRS allows two kinds of representations: graph and sequential notation (Figure 1). There are five types of semantic information involved in DRS: concepts (`read.v.01`, `paper.n.02`, `new.a.01`, ...), roles (`Agent`, `Theme`, `Time`, ...), constants (`speaker`, `hearer`, `now`, ...), comparison operators ($=$, $\prec$, $\sim$, ...) and discourse relations (`NEGATION`, `CONTINUATION`, `CONTRAST`, ...), where concepts and roles are represented by WordNet synsets (Fellbaum, 2000) and VerbNet thematic relations (Kipper et al., 2006) respectively.

2.2 DRS parsing

DRS parsing was originally applied to English and has been continuously extended to other Latin languages. Initially, rule-based systems were predominantly utilized by early parsers for analyzing small English texts (Johnson and Klein, 1986; Asher and Wada, 1988; Bos, 2004, 2008, 2015b). The first version of GMB (Basile et al., 2012) which pro-

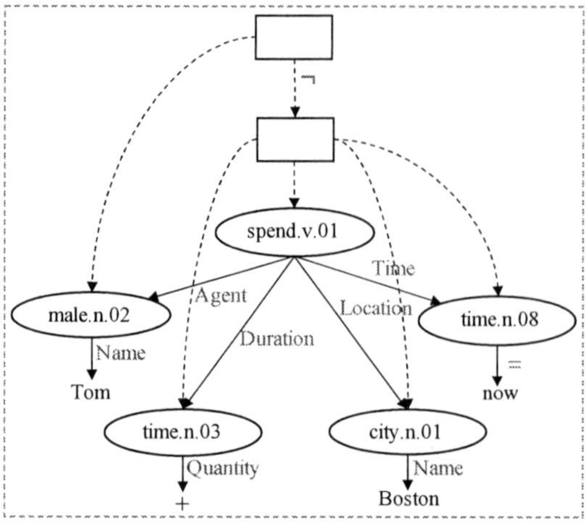

Figure 2: An example of a DRG with negation for sentence: "Tom doesn't spend much time in Boston."

vides English texts with DRS, is built on Boxer (Bos, 2008). With the release of PMB (Abzianidze et al., 2017) and the propose of the first shared tasks (Abzianidze et al., 2019), related research keeps growing, with a focus on deep learning models (Evang, 2019; Fancellu et al., 2019; van Noord et al., 2018, 2020a; Liu et al., 2019). The target languages have also expanded to other languages: German, Italian, Dutch and Chinese (Shen and Evang, 2022; Poelman et al., 2022a; Wang et al., 2021; Liu et al., 2021). Translation has been utilized in two manners when dealing with cross-lingual parsing: the first involves translating other languages into English and then employing an English parser, while the second involves translating English into other languages and training a parser specific to that language (Liu et al., 2021). In this paper, we use the existing Chinese-English parallel corpus to design a specific parser for Chinese, and compare the performance of the parser with the first method.

3 Data Creation

In previous work, for non-English parsing tasks, the semantic representation of English is usually directly used as the semantic representation of the target language, but most of these works focus on Latin languages (Fancellu et al., 2019; Ribeiro et al., 2021). For non-Latin languages such as Chinese, named entities are not language-neutral, as illustrated in the work of Wang et al. (2021), and are quite different from named entities in English texts. To design a more reasonable Chinese parser, we first focus on replacing the named entities in the English semantic representation with Chinese, so that the parser can parse out the Chinese named entities corresponding to the text content according to different texts.

To achieve our goal, we use the data of PMB, the largest parallel corpus of DRS data available, as our experimental object. From the PMB, English-Chinese parallel texts and DRS data for English texts are collected. Based on that, we propose a pipeline to obtain Chinese DRS for Chinese text. Our pipeline has three steps: (1) using tokenizers tools to segment Chinese and English text data; (2) utilizing the English-Chinese alignment tool to obtain the alignment tokens between Chinese and English texts; (3) replacing named entities in English DRS with Chinese named entities. Figure 3 shows our processing pipeline.

3.1 Text Tokenizers

Preprocessing data with a tokenizer is an important step in the pipeline because the alignment of Chinese and English texts needs to act on the data after tokenization. At the same time, since the quality of upstream results directly affects downstream performance, the quality of text segmentation also directly affects the correctness of Chinese and English text alignment. In this work, we use Moses (Koehn et al., 2007) for English, which is advanced and widely used. It is a collection of complex normalization and segmentation logic that works very well for structured languages like English. For Chinese, we choose HanLP (He and Choi, 2021), which is an efficient, user-friendly and extendable tokenizer. Different from a widely used Jieba tokenizer, HanLP is based on the CRF algorithm. It takes into account word frequency and context at the same time, and can better identify ambiguous words and unregistered words.

3.2 English-Chinese Alignment

In order to realize the replacement of named entities in English semantic representation with Chinese named entities, it is very important to obtain the correct alignment of Chinese and English texts, especially the alignment of named entities in the two texts. In order to quickly and effectively obtain the alignment data in Chinese and English, we choose the GIZA++ word aligning tool. GIZA++ is the most popular statistical alignment and MT toolkit (Och and Ney, 2000), which implements the lexical translation models of Brown et al. (1993) (IBM Models), and the Hidden-Markov

(a) Yunus founded the Grameen Bank 30 years ago .
NULL ({ 3 }) 尤努斯 ({ 1 }) 30 ({ 6 }) 年 ({ 7 }) 前 ({ 8 }) 创立 ({ 2 }) 了 ({ }) 格莱美 ({ 4 }) 银行 ({ 5 }) 。 ({ 9 })

(b) Yunus: 尤努斯
Grameen Bank: 格莱美银行

(c)

```
male.n.02       Name "Yunus"                      % Yunus  [1]
found.v.01      Agent -1 Theme +2 Time +5% founded  [2]
time.n.08       TPR now                           %
company.n.01    Name "Grameen Bank"               % the Grameen Bank  [3-5]
quantity.n.01   EQU 30                            % 30  [6]
year.n.01       Quantity -1 TAB now               % years  [7]
time.n.08       TIN -4 TAB -1                     % ago .  [8-9]
```

⇨

```
male.n.02       Name "尤努斯" "
found.v.01      Agent -1 Theme +2 Time +5
time.n.08       TPR now
company.n.01    Name "格莱美银行"
quantity.n.01   EQU 30
year.n.01       Quantity -1 TAB now
time.n.08       TIN -4 TAB -1
```

Figure 3: (a) Alignment tokens obtained by GIZA++ tool for English (*"Yunus founded the Grameen Bank 30 years ago."*) and Chinese ("尤努斯30年前创立了格莱美银行。"), (b) aligned named entities dictionary in above texts, (c) same meaning representations with different named entities for English text and Chinese text respectively.

alignment Model (Vogel et al., 1996), trained using expectation-maximization (EM). GIZA++ is highly effective at aligning frequent words in a corpus, but error-prone for infrequent words.

3.3 Replacing Named Entities

The last step to obtain the Chinese semantic representation is to replace the named entities in the English DRS with Chinese named entities. First, the English named entities in DRS data can be easily obtained according to the edge types between two nodes. When the edge type is `Name`, the output nodes are named entities in the DRG. After processing the Chinese and English texts with the GIZA++ tool in the second step, we can obtain alignment tokens between Chinese and English. On this basis, a named entity alignment dictionary can be obtained, and then the English named entities in the DRS data can be replaced with Chinese named entities based on this dictionary.

4 Methodology

4.1 Neural Models

We adopt Recurrent Neural Networks (RNN) equipped with Long Short-Term Memory units (LSTM; Hochreiter and Schmidhuber 1997) as our baseline models. Following the work of van Noord et al. (2020b), we use frozen mBERT (Devlin et al., 2019) embeddings to initialize the encoder. An attention-based LSTM architecture is used for the decoder, where the attention memory is the concatenation of the attention vectors among all the input tokens. In addition, the copy mechanism (Gu et al., 2016; Gulcehre et al., 2016) is added to the decoder, which can integrate the attention distribution into the final vocabulary distribution. The copy mechanism favors copying tokens from the source text into the target text instead of generating all target tokens only from the target vocabulary.

4.2 Evaluation

Given a document to the DRS parser, it will generate variable-free sequential notation DRS as shown in Figure 1(b). The evaluation tool for DRS parsing task was recently proposed by Poelman et al. (2022b) and is based on the AMR standard evaluation tool Smatch (Cai and Knight, 2013). By converting a sequential DRS into DRG, Penman notation format data (Kasper, 1989) can be obtained, as shown in Figure 4 (b), and then Smatch can be used to compute F-scores based on matching triples between system output and gold meanings.

However, we note that the scores given by the above evaluation tool have two flaws: (1) the evaluation scores are too inflated, and it is difficult to detect the differences between different parsers. (2) the evaluation tool only gives an overall score without evaluating the different types of constituent elements in the DRS, it is difficult to quantitatively determine what is the difficulty of the parser in the parsing process. Based on that, we propose to compress evaluation scores to improve the above evaluation methods and further propose fine-grained evaluation metrics for different subtasks according to different types of components in DRS.

4.2.1 Overall Evaluation

Our improvement strategy is mainly aimed at the representation of the Penman format of DRG. We mainly improve on two points, one is WordNet synsets representation, and the other is constants representation.

In the previous evaluation method, the WordNet

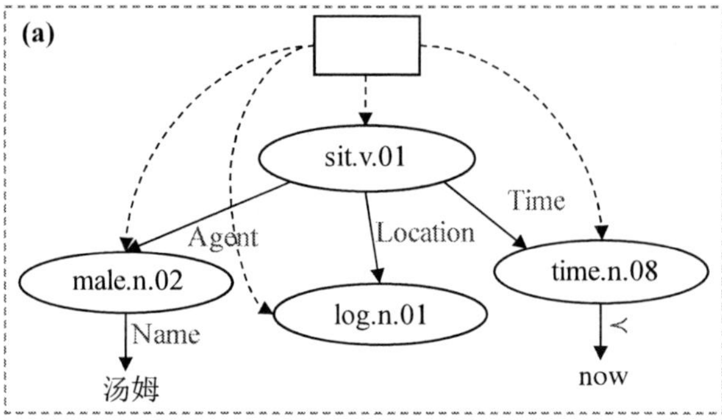


```
(b) ((b0 / "box"
        :member (s0 / "sense"
                :lemma "male"
                :pos "n"
                :sense "02"
                :Name (c0 / "汤姆"))
        :member (s1 / "sense"
                :lemma "sit"
                :pos "v"
                :sense "03"
                :Agent s0
                :Time (s2 / "sense"
                        :lemma "time"
                        :pos "n"
                        :sense "08"
                        :TPR (c1 / "now"))
                :Location (s3 / "sense"
                        :lemma "log"
                        :pos "n"
                        :sense "01"))
        :member s2
        :member s3)
```

```
(c) (b0 / "box"
        :member (s0 / "male.n.02"
                :Name "汤姆")
        :member (s1 / "sit.v.01"
                :Agent s0
                :Time (s2 / "time.n.08"
                        :TPR "now")
                :Location (s3 / "log.n.01"))
        :member s2
        :member s3)
```

Figure 4: (a) Graph structured DRS for Chinese sentence:" 汤姆坐在一根圆木上。". (b) Penman format of DRG with fine-grained WordNet synsets used for evaluation (Poelman et al., 2022b). (c) Penman format of DRG with coarse-grained WordNet synsets used for evaluation (Ours).

synsets in Penman format are fine-grained during the evaluation process, and the WordNet synsets are divided into three parts (lemma, pos, number) according to their constituents. On this basis, even if the parser generates wrong concepts, such as `time.n.08` and `time.n.01`, the Smatch still obtains a similar inflated F1 score. To this end, we change the WordNet synsets in the Penman format to a coarse-grained representation to strictly evaluate WordNet synsets qualities generated by parsers, as shown in Figure 4 (c). In addition, we have also modified the constant representation in Penmen format, such as the constant now shown in the figure, because the variable *c* is added to the constant, making the triples in Penman format redundant, which also makes the F1-score higher to a certain extent. By omitting the *c* variable as shown in Figure 4 (c), we further compress the F1-score.

4.2.2 Fine-grained Evaluation

To evaluate the quality of specific subtasks in DRS parsing, we imitate the fine-grained metrics for AMR parsing task (Damonte et al., 2017; Zhang et al., 2019) to DRS parsing. In order to make them compatible with DRS, we make some changes based on the data characteristics of DRS. Our fine-grained metrics consist of three categories in total: *graph-level*, *node-level* and *edge-level*. Each category includes more fine-grained evaluation metrics. All the metrics are proposed based on the semantic information types involved in DRS (see Section 2.1).

In ***graph-level*** evaluation, `No Roles`, `No Discourse`, `No Operators` and `No Senses` are used to represent the Smatch scores of the DRG in Penman format ignoring Roles, Discourse, Operators and Senses respectively. In theory, they are Smatch's coarse-grained scores, which are higher than the original Smatch scores.

In ***node-level*** evaluation, we compute F-score on the list of parsed information types (such as roles, constants, and discourse relations) instead of using Smatch. Note that different from the metrics in the AMR parsing task, concepts in DRS are represented by WordNet synsets, so `Concepts` can be evaluated more finely by part-of-speech (`noun`, `adjective`, `adverb` and `verb`). `Discourse` detects all discourse relation labels except *NEGATION* since it is more common and specific in DRS than other discourse relations labels, the `Negation` metric is used for evaluation to detect *NEGATION* edge label alone. In addition, `Member` metric is added to evaluate the ratio of the generated concepts. In DRG, *member* represents the edge label

Alignment Error	Data Type	Example
Dislocation	Chinese	梅尔·卡玛津 是 天狼星 的 执行官。
	English	Mel Karmazin is an executive of Sirius .
	Wrong	male.n.02 Name "梅尔·卡玛津" be.v.08 Theme -1 Time +1 Co-Theme +2 time.n.08 EQU now person.n.01 Role +1 executive.n.01 Of +1 company.n.01 Name "执行官"
	Corrected	male.n.02 Name "梅尔·卡玛津" be.v.08 Theme -1 Time +1 Co-Theme +2 time.n.08 EQU now person.n.01 Role +1 executive.n.01 Of +1 company.n.01 Name "天狼星"
Character Exclusion	Chinese	什么 乐队 唱 了 " 快乐 在 一起 " 这 首 歌？
	English	What group sang the song " Happy Together " ?
	Wrong	group.n.01 Name ? sing.v.02 Agent -1 Time +1 Theme +2 time.n.08 TPR now song.n.01 EQU +1 music.n.01 Name "快乐一起"
	Corrected	group.n.01 Name ? sing.v.02 Agent -1 Time +1 Theme +2 time.n.08 TPR now song.n.01 EQU +1 music.n.01 Name "快乐在一起"
Character Inclusion	Chinese	卢瑟福·海斯 1822 年 出生 于俄亥俄州 。
	English	Rutherford Hayes was born in Ohio in 1822 .
	Wrong	male.n.02 Name "卢瑟福·海斯1822" time.n.08 TPR now bear.v.02 Patient -2 Location +1 Time +2 state.n.01 Name "俄亥俄州" time.n.08 YearOfCentury 1822 TIN -3
	Corrected	male.n.02 Name "卢瑟福·海斯" time.n.08 TPR now bear.v.02 Patient -2 Location +1 Time +2 state.n.01 Name "俄亥俄州" time.n.08 YearOfCentury 1822 TIN -3
Nationality	Chinese	我 不 是 爱尔兰人 。
	English	I am not Irish .
	Wrong	person.n.01 EQU speaker NEGATION <1 time.n.08 EQU now be.v.03 Theme -2 Time -1 Source +1 country.n.02 Name ""
	Corrected	person.n.01 EQU speaker NEGATION <1 time.n.08 EQU now be.v.03 Theme -2 Time -1 Source +1 country.n.02 Name "ireland"

Table 1: Alignment errors illustrated by four examples. In Chinese and English texts, words of the same color indicate correct alignment between them. Inbformation marked in red is the wrong named entity obtained by the GIZA++ tool. Text in green indicates the correct named entity in the corrected DRS.

connecting the *BOX* node and the concepts node, i.e., the dashed line as shown in Figure 4 (a).

For ***edge-level*** evaluation, we focus on calculating the F-score based on the number of matching triples in the parsed DRG and the gold DRG. For example, `Names` in edge-level is a metric that considers the relations between concepts nodes and named entities, which differs from the metric of `Names` in node-level, which only considers the concepts labeled with *Name* and ignores the accuracy of named entities themselves. [1]

5 Experiments

5.1 Dataset

We collect all Chinese-English text pairs in the PMB. According to the quality label of English DRS, we divide the data into gold data and silver data, and randomly split the test set and development set from the gold data. Since PMB data may contain duplicate data, before splitting, we first filter the duplicate data. Then we merge the remaining gold data and silver data as our training set, and get a total of 137,781 training instances, 1,000 development instances and 1,000 test instances, each instance contains English DRS data, corresponding English text, and Chinese text. [2]

After splitting the data, we use the pipeline introduced in Section 3 to process our Chinese and English texts to get the Chinese and English word alignment data, and then replace the named entities in the English DRS with Chinese. However, we noticed that not all replacements were successful. We classified the wrong replacement types into four types, as shown in Table 1. These errors are mainly caused by GIZA++ alignment errors when aligning Chinese and English text words. Among them, the fourth type of error is quite special. In our experiment, we directly ignore the location named entities used to refer to nationality and do not replace them with Chinese named entites. In order to reduce the work of manual correction and make the work reproducible, We only fix incorrect named entity replacements in the test set, where 26 of the

[1]Our evaluation suite is available at: `https://github.com/wangchunliu/SBN-evaluation-tool`.

[2]Our data and code are available at: `https://github.com/wangchunliu/Chinese-SBN-parsing`.

Alignment Error	Reason	Example
Named-entities	Jieba	English: Melanie killed a spider with her hand . Chinese: 媚兰用 (1 5 6) 手 (7) 杀死 (2) 了 ( ) 一只 (3) 蜘蛛 (4) 。 (8)
	HanLP	English: Melanie killed a spider with her hand . Chinese: 媚兰 (1 6) 用 (5) 手 (7) 杀死 (2) 了 ( ) 一 (3) 只 ( ) 蜘蛛 (4) 。 (8)
Information units	gold data	The ground floor was flooded . Chinese: 一楼 (1) 被 ( ) 淹 (2 3 4 5) 了 ( ) 。 (6)
	all data	English: The ground floor was flooded . Chinese: 一楼 (1 2 3) 被 (4) 淹 (5) 了 ( ) 。 (6)

Table 2: Impact of different tokenizers and data sizes on GIZA++ performance.

1000 test set instances require manual correction of named entities.

5.2 Settings

For tokenizers, we use Moses (Koehn et al., 2007) and HanLP (He and Choi, 2021) on English and Chinese respectively. We observe that the HanLP tokenizer outperforms Jieba[3], a tokenizer widely used in Chinese, in segmenting text containing named entities. This is an important indicator for selecting a tokenizer, because getting the correct Chinese and English named entity pairs is our main goal. In addition, we observed that HanLP's segmentation results also outperformed Jieba's tokenizer on text containing traditional Chinese characters, while the Chinese data in PMB contains traditional Chinese characters. This is also one of the reasons for choosing the HanLP tokenizer. At the top of Table 2, we show the difference in name entities between the Jieba tokenizer and the HanLP tokenizer. In addition, we give an example of the impact of different sizes of training data on the alignment performance of GIZA++ at the bottom of Table 2, and the results show that it is almost impossible to achieve correct alignment using only gold data.

	Document-level			Word-level	
Data	**Train**	**dev**	**test**	**src**	**tgt**
English	137,781	1,000	1,000	38,441	39,761
Chinese	137,781	1,000	1,000	42,446	41,734

Table 3: Document statistics and vocabulary sizes.

All experiments are implemented based on OpenNMT (Klein et al., 2017). For the vocabulary, we construct vocabularies from all words, the vocabulary sizes as shown in Table 3. The hyperparameters are set based on performance on the development set. We use SGD optimizer with the initial learning rate set to 1 and decay 0.8. In addition, we set the dropout to 0.5 at the decoder layer to avoid overfitting with batch size 32.

[3]https://github.com/fxsjy/jieba

Metric	EN	ZH	ZH→EN$_{zh}$
Smatch$_1$	91.0	86.0	84.7
Smatch$_2$	88.9	83.8	81.7
Well-formed	99.8	99.7	99.7
Graph-level			
No Roles	90.0	85.5	84.2
No Discourse	89.5	83.9	82.7
No Operators	89.5	84.7	83.4
No Senses	91.9	85.6	84.7

Table 4: F-scores with Smatch on the test set of semantic parsers. Note: Smatch$_1$ and Smatch$_2$ represent the original evaluation (Poelman et al., 2022b) and our improved evaluation.

5.3 Main Results

Table 4 shows the results obtained by the parsers with Smatch, which gives the overall performance for different parsers. The first parser (**EN**) is trained on the English dataset based on the model introduced in Section 4.1. The Smatch$_1$ result of our English parser is slightly lower than the results of Poelman et al. (2022b), which we believe is due to slightly different training, development and test set instances. The result of Smatch$_2$ is significantly lower than the result of Smatch$_1$, indicating that the F1-score has been significantly compressed and will not be too inflated (see Section 4).

The Chinese parser (**ZH**) is trained on the data created by the pipeline introduced in Section 3. The results show that the performance of the Chinese parser is lower than the English parser in all overall evaluation metrics. **ZH→EN$_{zh}$** shows the perfor-

	Metric	EN	ZH	ZH→EN_{zh}
Node	Names	70.8	66.0	67.7
	Negation	92.3	88.7	88.8
	Discourse	86.0	80.4	75.2
	Roles	89.2	84.0	84.9
	Members	97.5	95.4	95.9
	Concepts	81.2	73.3	74.4
	noun	87.1	82.1	83.3
	adj	73.3	54.2	52.5
	adv	76.8	35.3	45.5
	verb	59.7	45.5	47.2
Edge	Roles	81.0	73.3	73.7
	Names	79.4	74.0	45.5
	Members	90.9	86.4	87.0
	Operators	92.9	87.7	87.7
	Discourse	86.2	79.6	75.3

Table 5: F-scores of fine-grained evaluation on the test set of semantic parsers. The evaluation metrics in the table are all based on the Penman format DRG with coarse-grained WordNet synsets.

mance by using the English parser on English text translated from Chinese text instead of training a dedicated model for Chinese text. The only unreasonable point is that the model will generate English named entities, which may not be recognized as the correct Chinese semantic representation.

The $smatch_1$ scores and the $smatch_2$ scores show that the Chinese parser outperforms using the ZH→EN_{zh} approach. For the metrics `No Senses` and `No Roles`, the evaluation results have been significantly improved compared with $Smatch_2$. This shows that Concepts and Roles have a greater impact on evaluation results than Discourse and Operators. It is worth noting that the performance difference between the Chinese and English parsers is about five percentage points across all metrics, while the difference between the ZH and the ZH→EN_{zh} narrows at the graph-level metrics compared to $Smatch_2$ score.

5.4 Fine-grained Results and Analysis

To further explore the performance of parsers, we apply our proposed fine-grained evaluation metrics to the results of two parsers. Tabel 5 shows the fine-grained evaluation performance of different component types based on DRG at node-level and edge-level.

`Names:` From the results, we observe that the metric `Names` gives completely opposite results at different evaluation levels. On the node-level, the `Names` metric in ZH parser scores the lowest, but on the edge-level, `Names` metric in ZH→EN_{zh} gives the lowest scores. This is reasonable and expected because the node-level `Names` metric only evaluates whether the parser can parse concepts to contain named entities, so the results of ZH→EN_{zh} parser should be similar to those of the English parser. However, the edge-level `Names` metric evaluates whether the generated named entities completely match the original text, and the ZH→EN_{zh} parser completely loses the Chinese named entity information.

`Discourse:` An important observation is that the metric `Discourse` has very low F1 scores on both the node-level and the edge-level for the Chinese parser. Using machine translation and an English parser to parse Chinese (ZH→EN_{zh}) will further degrade the performance of the metric `Discourse`. Based on the text data and parsed output, we find that discourse relations in Chinese are inconspicuous, and even disappears after being translated into English (see Table 6 for examples).

`Concepts:` Table 5 shows the `Concepts` scores of ZH parser are lower than those for ZH→EN_{zh} except for the *adj* category. This is an interesting finding, because the performance of other parts of speech in the ZH parser is worse than that of ZH→EN_{zh}, while *adj* is special. We observe that the expressions of adjectives in Chinese translated into English are diverse and may not match the original English text (see Table 6 and Appendix B for relevant examples).

For the English parser, *verbs* are the most difficult words to parse, scoring significantly lower than other parts of speech. However, the difficulty of Chinese semantic parsing is mainly reflected in *adv*. In addition, the accuracy of ZH→EN_{zh} in parsing concepts of *adv* is significantly better than that of the ZH parser, but it is still the lowest results in four types of parts of speech for ZH→EN_{zh}. On the one hand, the corpus containing adverb data is smaller, which makes the training insufficient. On the other hand, the adverbs in Chinese are usually not obvious and diverse.

For *noun* and *verb*, ZH has the worst performance, with the ZH→EN_{zh} method, the performance of *noun* and *verb* is slightly improved, but it is much worse than the **EN** parser. A typical reason is that the English text translated from Chinese may not be consistent with the original English text. We

Information Type	Example	Lost/Changed in Translation
Discourse	EN: A parrot can mimic a person's voice. ZH: 鹦鹉会模仿人的声音。 ZH→EN: Parrots mimic human voices.	POSSIBILITY Lost
	EN: Tom asks his mother if she can buy him a new toy. ZH: 汤姆请求他母亲给他买新玩具。 ZH→EN: Tom begged his mother to buy him new toys.	ATTRIBUTION Lost
Concepts	EN: That guy is completely nuts! ZH: 那家伙真是疯了！ ZH→EN: That guy is crazy!	Adverb Lost
	EN: She's very handy with a saw. ZH: 她很会用锯子。 ZH→EN: She is good with a saw.	Adjective Changed
	EN: I'm awake. ZH: 我醒了。 ZH→EN: I woke up.	Adjective Lost
	EN: Tom is suffering from a bad headache. ZH: 汤姆头痛得厉害。 ZH→EN: Tom has a bad headache.	Verb Changed
Operators	EN: I slept on the boat. ZH: 我睡在船上。 ZH→EN: I sleep on the boat.	Tense Lost
Negation	EN: The music lured everyone. ZH: 音乐吸引了所有人。 ZH→EN: Music appeals to all.	NEGATION Lost
	EN: The printer doesn't work. ZH: 打印机坏了。 ZH→EN:The printer is broken.	

Table 6: Examples of translated English texts with loss of information.

observe that the DRS sequences parsed using the translated text are overall shorter than those parsed using the original English text, some noun concepts are missing, and the verb concepts may be inconsistent with the reference DRS (see Appendix B for examples).

Operators & Negation: Our fine-grained results obtained by using machine translation and the English parser are not always worse than training a Chinese parser alone. For the metrics Negation and Operators, both methods have similar scores at both the node-level and the edge-level. However, when we compare the results of ZH→EN_{zh} with EN parser, we find that all the results of ZH→EN_{zh} are significantly lower than those of the EN parser. We found that tense information is usually lost in the process of English-Chinese translation, but almost no tense information is lost in the process of Chinese-English translation. This explains why the result of the Chinese parser operator is significantly lower than that of the English parser, while the result of ZH→EN_{zh} is the same as that of the ZH parser. For Negation, we can observe something interesting. As the connector NEGATION in English DRss can also express universal quantification (using nesting of two negation operators) for words such as "every" and "always", this information is missing in the translation process, and as a result not picked up by the parser.

Members & Roles: For this metric, ZH→EN_{zh} even slightly outperforms the ZH parser, but they are both lower than the EN parser. On the one hand, a free translation may lead to a different ordering of semantic information. Although texts with the same meaning but realised with different word order have the same semantic graph, a parser based on sequence-to-sequence neural networks may get the wrong graph structure leading to a lower evaluation score of the Roles evaluation metric. On the other hand, both evaluation metrics are affected by the correctness of Concepts, and in our results, the Chinese parser scored lower than the other two parsers for Concepts.

6 Conclusion

Given an annotated meaning bank primarily designed for English, it is feasible to develop a semantic parser for Chinese by pairing the "English" meaning representation with Chinese translations, reaching good results. Most difficulties in Chinese parsing are caused by adverbs, while the diversity of Chinese verbs and adjectives also has a big impact on parsing performance. Using Machine Translation as an alternative to approach semantic parsing for Chinese yields slightly lower results. Our fine-grained graph evaluation gives better insight when comparing different parsing approaches.

Acknowledgments

This work was funded by the NWO-VICI grant "Lost in Translation—Found in Meaning" (288-89-003) and the China Scholarship Council (CSC). We thank the anonymous reviewers for detailed comments that improved this paper. We would also like to thank the Center for Information Technology of the University of Groningen for their support and for providing access to the Peregrine high performance computing cluster.

References

Lasha Abzianidze, Johannes Bjerva, Kilian Evang, Hessel Haagsma, Rik van Noord, Pierre Ludmann, Duc-Duy Nguyen, and Johan Bos. 2017. The Parallel Meaning Bank: Towards a multilingual corpus of translations annotated with compositional meaning representations. In *Proceedings of the 15th Conference of the European Chapter of the Association for Computational Linguistics: Volume 2, Short Papers*, pages 242–247, Valencia, Spain. Association for Computational Linguistics.

Lasha Abzianidze, Rik van Noord, Hessel Haagsma, and Johan Bos. 2019. The first shared task on discourse representation structure parsing. In *Proceedings of the IWCS Shared Task on Semantic Parsing*, Gothenburg, Sweden. Association for Computational Linguistics.

Nicholas Asher and Haijime Wada. 1988. A Computational Account of Syntactic, Semantic and Discourse Principles for Anaphora Resolution. *Journal of Semantics*, 6(1):309–344.

Valerio Basile, Johan Bos, Kilian Evang, and Noortje Venhuizen. 2012. A platform for collaborative semantic annotation. In *Proceedings of the Demonstrations at the 13th Conference of the European Chapter of the Association for Computational Linguistics (EACL)*, pages 92–96, Avignon, France.

Michele Bevilacqua, Rexhina Blloshmi, and Roberto Navigli. 2021. One SPRING to rule them both: Symmetric AMR semantic parsing and generation without a complex pipeline. In *Proceedings of AAAI*.

Johan Bos. 2004. Computational semantics in discourse: Underspecification, resolution, and inference. *Journal of Logic, Language and Information*, 13:139–157.

Johan Bos. 2008. Wide-coverage semantic analysis with Boxer. In *Semantics in Text Processing. STEP 2008 Conference Proceedings*, pages 277–286. College Publications.

Johan Bos. 2015a. Open-domain semantic parsing with boxer. In *Proceedings of the 20th Nordic Conference of Computational Linguistics, NODALIDA 2015, May 11-13, 2015, Institute of the Lithuanian Language, Vilnius, Lithuania*, pages 301–304. Linköping University Electronic Press / ACL.

Johan Bos. 2015b. Open-domain semantic parsing with boxer. In *Nordic Conference of Computational Linguistics*.

Johan Bos. 2021. Variable-free discourse representation structures. *Semantics Archive*.

Peter F. Brown, Stephen A. Della Pietra, Vincent J. Della Pietra, and Robert L. Mercer. 1993. The mathematics of statistical machine translation: Parameter estimation. *Computational Linguistics*, 19(2):263–311.

Shu Cai and Kevin Knight. 2013. Smatch: an evaluation metric for semantic feature structures. In *Proceedings of the 51st Annual Meeting of the Association for Computational Linguistics (Volume 2: Short Papers)*, pages 748–752, Sofia, Bulgaria. Association for Computational Linguistics.

Marco Damonte, Shay B. Cohen, and Giorgio Satta. 2017. An incremental parser for Abstract Meaning Representation. In *Proceedings of the 15th Conference of the European Chapter of the Association for Computational Linguistics: Volume 1, Long Papers*, pages 536–546, Valencia, Spain. Association for Computational Linguistics.

Jacob Devlin, Ming-Wei Chang, Kenton Lee, and Kristina Toutanova. 2019. BERT: Pre-training of deep bidirectional transformers for language understanding. In *Proceedings of the 2019 Conference of the North American Chapter of the Association for Computational Linguistics: Human Language Technologies, Volume 1 (Long and Short Papers)*, pages 4171–4186, Minneapolis, Minnesota. Association for Computational Linguistics.

Kilian Evang. 2019. Transition-based DRS parsing using stack-LSTMs. In *Proceedings of the IWCS Shared Task on Semantic Parsing*, Gothenburg, Sweden. Association for Computational Linguistics.

Federico Fancellu, Sorcha Gilroy, Adam Lopez, and Mirella Lapata. 2019. Semantic graph parsing with recurrent neural network DAG grammars. In *Proceedings of the 2019 Conference on Empirical Methods in Natural Language Processing and the 9th International Joint Conference on Natural Language Processing (EMNLP-IJCNLP)*, pages 2769–2778, Hong Kong, China. Association for Computational Linguistics.

Christiane D. Fellbaum. 2000. Wordnet : an electronic lexical database. *Language*, 76:706.

Jiatao Gu, Zhengdong Lu, Hang Li, and Victor O.K. Li. 2016. Incorporating copying mechanism in sequence-to-sequence learning. In *Proceedings of the 54th Annual Meeting of the Association for Computational Linguistics (Volume 1: Long Papers)*, pages 1631–1640, Berlin, Germany. Association for Computational Linguistics.

Caglar Gulcehre, Sungjin Ahn, Ramesh Nallapati, Bowen Zhou, and Yoshua Bengio. 2016. Pointing the unknown words. In *Proceedings of the 54th Annual Meeting of the Association for Computational Linguistics (Volume 1: Long Papers)*, pages 140–149, Berlin, Germany. Association for Computational Linguistics.

Han He and Jinho D. Choi. 2021. The stem cell hypothesis: Dilemma behind multi-task learning with transformer encoders. In *Proceedings of the 2021 Conference on Empirical Methods in Natural Language Processing*, pages 5555–5577, Online and Punta Cana, Dominican Republic. Association for Computational Linguistics.

Sepp Hochreiter and Jürgen Schmidhuber. 1997. Long short-term memory. *Neural Comput.*, 9(8):1735–1780.

Matic Horvat, Ann Copestake, and Bill Byrne. 2015. Hierarchical statistical semantic realization for Minimal Recursion Semantics. In *Proceedings of the 11th International Conference on Computational Semantics*, pages 107–117, London, UK. Association for Computational Linguistics.

Mark Johnson and Ewan Klein. 1986. Discourse, anaphora and parsing. In *Proceedings of the 11th Coference on Computational Linguistics*, COLING '86, page 669–675, USA. Association for Computational Linguistics.

Hans Kamp and U. Reyle. 1993. From discourse to logic: Introduction to model theoretic semantics of natural language, formal logic and discourse representation theory. *Language*, 71(4).

Robert T. Kasper. 1989. A flexible interface for linking applications to Penman's sentence generator. In *Speech and Natural Language: Proceedings of a Workshop Held at Philadelphia, Pennsylvania, February 21-23, 1989*.

Karin Kipper, Anna Korhonen, Neville Ryant, and Martha Palmer. 2006. Extending VerbNet with novel verb classes. In *Proceedings of the Fifth International Conference on Language Resources and Evaluation (LREC'06)*, Genoa, Italy. European Language Resources Association (ELRA).

Guillaume Klein, Yoon Kim, Yuntian Deng, Jean Senellart, and Alexander Rush. 2017. OpenNMT: Open-source toolkit for neural machine translation. In *Proceedings of ACL 2017, System Demonstrations*, pages 67–72, Vancouver, Canada. Association for Computational Linguistics.

Philipp Koehn, Hieu Hoang, Alexandra Birch, Chris Callison-Burch, Marcello Federico, Nicola Bertoldi, Brooke Cowan, Wade Shen, Christine Moran, Richard Zens, Chris Dyer, Ondřej Bojar, Alexandra Constantin, and Evan Herbst. 2007. Moses: Open source toolkit for statistical machine translation. In *Proceedings of the 45th Annual Meeting of the Association for Computational Linguistics Companion Volume Proceedings of the Demo and Poster Sessions*, pages 177–180, Prague, Czech Republic. Association for Computational Linguistics.

Jiangming Liu, Shay B. Cohen, and Mirella Lapata. 2019. Discourse representation structure parsing with recurrent neural networks and the transformer model. In *Proceedings of the IWCS Shared Task on Semantic Parsing*, Gothenburg, Sweden. Association for Computational Linguistics.

Jiangming Liu, Shay B. Cohen, Mirella Lapata, and Johan Bos. 2021. Universal Discourse Representation Structure Parsing. *Computational Linguistics*, 47(2):445–476.

Franz Josef Och and Hermann Ney. 2000. Improved statistical alignment models. In *Proceedings of the 38th Annual Meeting on Association for Computational Linguistics*, ACL '00, page 440–447, USA. Association for Computational Linguistics.

Franz Josef Och and Hermann Ney. 2003. A systematic comparison of various statistical alignment models. *Computational Linguistics*, 29(1):19–51.

Wessel Poelman, Rik van Noord, and Johan Bos. 2022a. Transparent semantic parsing with Universal Dependencies using graph transformations. In *Proceedings of the 29th International Conference on Computational Linguistics*, pages 4186–4192, Gyeongju, Republic of Korea. International Committee on Computational Linguistics.

Wessel Poelman, Rik van Noord, and Johan Bos. 2022b. Transparent semantic parsing with Universal Dependencies using graph transformations. In *Proceedings of the 29th International Conference on Computational Linguistics*, pages 4186–4192, Gyeongju, Republic of Korea. International Committee on Computational Linguistics.

Leonardo F. R. Ribeiro, Jonas Pfeiffer, Yue Zhang, and Iryna Gurevych. 2021. Smelting gold and silver for improved multilingual AMR-to-Text generation. In *Proceedings of the 2021 Conference on Empirical Methods in Natural Language Processing*, pages 742–750, Online and Punta Cana, Dominican Republic. Association for Computational Linguistics.

Minxing Shen and Kilian Evang. 2022. DRS parsing as sequence labeling. In *Proceedings of the 11th Joint Conference on Lexical and Computational Semantics*, pages 213–225, Seattle, Washington. Association for Computational Linguistics.

Rik van Noord, Lasha Abzianidze, Antonio Toral, and Johan Bos. 2018. Exploring neural methods for parsing discourse representation structures. *Transactions of the Association for Computational Linguistics*, 6:619–633.

Rik van Noord, Antonio Toral, and Johan Bos. 2019. Linguistic information in neural semantic parsing with multiple encoders. In *Proceedings of the 13th International Conference on Computational Semantics*

- *Short Papers*, pages 24–31, Gothenburg, Sweden. Association for Computational Linguistics.

Rik van Noord, Antonio Toral, and Johan Bos. 2020a. Character-level representations improve DRS-based semantic parsing even in the age of BERT. In *Proceedings of the 2020 Conference on Empirical Methods in Natural Language Processing (EMNLP)*, pages 4587–4603, Online. Association for Computational Linguistics.

Rik van Noord, Antonio Toral, and Johan Bos. 2020b. Character-level representations improve DRS-based semantic parsing even in the age of BERT. In *Proceedings of the 2020 Conference on Empirical Methods in Natural Language Processing (EMNLP)*, pages 4587–4603, Online. Association for Computational Linguistics.

Stephan Vogel, Hermann Ney, and Christoph Tillmann. 1996. HMM-based word alignment in statistical translation. In *COLING 1996 Volume 2: The 16th International Conference on Computational Linguistics*.

Chunliu Wang, Rik van Noord, Arianna Bisazza, and Johan Bos. 2021. Input representations for parsing discourse representation structures: Comparing English with Chinese. In *ACL/IJCNLP (2)*, pages 767–775. Association for Computational Linguistics.

Tianming Wang, Xiaojun Wan, and Hanqi Jin. 2020. AMR-to-text generation with graph transformer. *Transactions of the Association for Computational Linguistics*, 8:19–33.

Sheng Zhang, Xutai Ma, Kevin Duh, and Benjamin Van Durme. 2019. AMR parsing as sequence-to-graph transduction. In *Proceedings of the 57th Annual Meeting of the Association for Computational Linguistics*, pages 80–94, Florence, Italy. Association for Computational Linguistics.

A Result Plots

According to the fine-grained evaluation results, for both English and Chinese DRS parsing, relatively low f1 scores tend to appear in `Names` and `Concepts`. The performance of parser declined by approximately five percent after the named entity was converted to Chinese, especially the `adj` and `adv`, comparing **EN** with **ZH**.

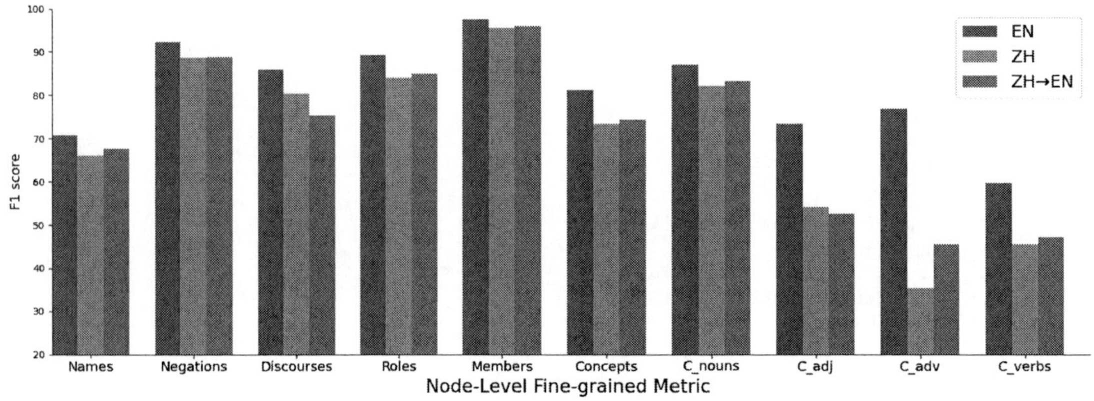

Figure 5: Fine-grained results among **EN**, **ZH**, and **ZH→EN**$_{zh}$ in Node-level.

B Output DRS

Number	Type	Example
No.1	EN Text	The music **lured everyone**.
	ZH Text	音乐吸引了所有人。
	ZH→EN	Music appeals to all.
	EN	music.n.01 NEGATION <1 person.n.01 NEGATION <1 surprise.v.02 Stimulus -2 Experiencer -1 Time +1 time.n.08 EQU now
	ZH	music.n.01 NEGATION <1 person.n.01 NEGATION <1 appeal.v.01 Agent -2 Theme -1 Time +1 time.n.08 TPR now
	ZH→EN$_{zh}$	event.v.01 Participant +1 music.n.01 appeal.v.01 Theme -1
	Gold DRS	music.n.01 NEGATION <1 person.n.01 NEGATION <1 lure.v.01 Agent -2 Patient -1 Time +1 time.n.08 TPR now
No.2	EN Text	She's **very handy** with a saw.
	ZH Text	她很会用锯子。
	ZH→EN	She is good with a saw.
	EN	female.n.02 time.n.08 EQU now very.r.01 handy.a.01 AttributeOf -3 Time -2 Degree -1 Instrument +1 saw.n.02
	ZH	female.n.02 time.n.08 TSU now use.v.01 Agent -2 Time -1 Theme +1 Instrument +2 entity.n.01 saw.n.02
	ZH→EN$_{zh}$	female.n.02 time.n.08 EQU now good.a.01 AttributeOf -2 Time -1 Instrument +1 saw.n.02
	Gold DRS	female.n.02 time.n.08 EQU now very.r.01 handy.a.03 AttributeOf -3 Time -2 Degree -1 Instrument +1 saw.n.02

Table 7: Examples of output DRSs by different parsers.

Author Index

Association for Computational Linguistics
209 N. Eighth Street
Stroudsburg, Pennsylvania 18360

ISBN 978-1-7138-9057-7

CURRAN ASSOCIATES INC.
proceedings
.com

Transactions of the Association for Computational Linguistics

(TACL 2013)

TACL Volume 1, Issue 1

ISBN: 978-1-62748-394-0